Rapid Assessment Program

21

RAP Bulletin of Biological Assessment

Southern New Ireland, Papua New Guinea: A Biodiversity Assessment

Bruce M. Beehler and
Leeanne E. Alonso, Editors

Center for Applied Biodiversity Science (CABS)

Conservation International

Department of Environment and Conservation, Papua New Guinea

Papua New Guinea National Museum and Art Gallery

Bernice P. Bishop Museum, USA

National Museum of Natural History, USA

The RAP Bulletin of Biological Assessment is published by:
Conservation International
Center for Applied Biodiversity Science
Department of Conservation Biology
1919 M St. NW, Suite 600
Washington, DC 20036
USA

202-912-1000 telephone
202-912-9773 fax
www.conservation.org
www.biodiversityscience.org

Editors: Bruce M. Beehler and Leeanne E. Alonso
Design/Production: Kim Meek
Maps: Bruce M. Beehler
Cover Photographs: Bruce M. Beehler

Conservation International is a private, non-profit organization exempt from federal income tax under section 501c(3) of the Internal Revenue Code.

ISBN 1-881173-61-5

Library of Congress Catalog Card Number: 2001099176

RAP Bulletin of Biological Assessment was formerly RAP Working Papers. Numbers 1–13 of this series were published under the previous title.

Suggested citation:
Beehler, B. M. and L. E. Alonso (Editors). 2001. Southern New Ireland, Papua New Guinea: A Biodiversity Assessment. RAP Bulletin of Biological Assessment 21. Conservation International, Washington, DC.

Funding for this study was provided by a USAID grant through the Biodiversity Support Program (BSP/WWF Agreement No. ME26, USAID #DHR 5554 A 00 8044 00). Publication of this report was funded by the generous support of The Giuliani Family Foundation.

Using New Leaf Opaque 60# smooth text paper (80% recycled/60% post-consumer waste), and bleached without the use of chlorine or chlorine compounds results in measurable environmental benefits[1]. For this report, using 1,404 pounds of post-consumer waste instead of virgin fiber saved…

5 Trees
457 Pounds of solid waste
502 Gallons of water
655 Kilowatt hours of electricity (equal to .8 months of electric power required by the average U.S. home)
830 Pounds of greenhouse gases (equal to 672 miles travelled in the average American car)
4 Pounds of HAPs, VOCs, and AOX combined
1 Cubic yard of landfill space

[1] Environmental benefits are calculated based on research done by the Environmental Defense Fund and the other members of the Paper Task Force who studied the environmental impacts of the paper industry. Contact the EDF for a copy of their report and the latest updates on their data. Trees saved calculation based on trees with a 10" diameter. Actual diameter of trees cut for pulp range from 6" up to very large, old growth trees. Home energy use equivalent provided by Pacific Gas and Electric Co., San Francisco. Hazardous Air Pollutants (HAPs), Volatile Organic Compounds (VOCs), and Absorbable Organic Compounds (AOX). Landfill space saved based on American Paper Institute, Inc. publication, *Paper Recycling and its Role in Solid Waste Management.*

Table of Contents

Appendices

Participants and Contributors

Institutions given indicate affiliations at the time of the RAP expedition (1994).
*Current affiliation.
Email addresses are current at time of publication (2001).

Research Team

Bruce M. Beehler
Ornithologist, RAP Team Leader
*Conservation International
b.beehler@conservation.org

Robin B. Foster
Plant Ecologist
Conservation International
*Field Museum, Chicago, USA
rfoster@fmnh.org

Joseph Wiakabu
Botanist
PNG Forest Research Institute
Lae, Papua New Guinea

Louise H. Emmons
Mammalogist
Conservation International
*National Museum of Natural History, USA
emmons.louise@nmnh.si.edu

Felix Kinbag
Mammalogist
PNG Department of Environment and Conservation

Allen Allison
Herpetologist
Bernice P. Bishop Museum, Hawaii
allison@hawaii.edu

Ilaiah Bigilale
Vertebrate Biologist
PNG National Museum and Art Gallery

Wayne Takeuchi
Botanist
PNG Forest Research Institute
wtakeuchi@global.net.pg

Larry Orsak
Entomologist
*Scientific Methods, Inc., USA
lesmangi@hotmail.com

Tommy Kosi
Entomologist
PNG Forest Research Institute

J. Phillip Angle
Vertebrate Biologist
National Museum of Natural History, USA

Daink Kuro
Ornithologist
PNG Department of Environment and Conservation

Trip Coordinators

Mr. Samuel Antiko
Project Leader
Department of Environment and Conservation

Michael Hedemark
Camp Coordinator and Ornithologist
Conservation Resource Centre
hedemark@laonet.net

Rob McCallum
Field Site Coordinator
Conservation Resource Centre

Other Scholarly Contributors

Daniel Polhemus
Entomologist
Bernice P. Bishop Museum, Hawaii, USA
*National Museum of Natural History, USA

Nicolas Eason
Entomologist
Christensen Research Institute, PNG

David Gibbs
Consultant Ecologist
Maidenhead, England

Editors

Bruce M. Beehler
Conservation International, USA
b.beehler@conservation.org

Leeanne E. Alonso
Conservation International, USA
l.alonso@conservation.org

Organizational Profiles

Conservation International (CI)

Conservation International (CI) is an international, nonprofit organization based in Washington, DC. CI acts on the belief that the Earth's natural heritage must be maintained if future generations are to thrive spiritually, culturally, and economically. Our mission is to conserve biological diversity and the ecological processes that support life on Earth, and to demonstrate that human societies are able to live harmoniously with nature.

Conservation International
1919 M Street, NW, Suite 600
Washington, DC 20036 USA
(tel) 202 912-1000
(fax) 202 912-0773
www.conservation.org
www.biodiversityscience.org

Conservation International—Papua New Guinea Program

P. O. Box 106
Waigani, NCD
Papua New Guinea
(tel) 675 323-1532
(fax) 675 325-4234
cimoresby@dg.com.pg

Department of Environment and Conservation (DEC), Papua New Guinea

The DEC is vested with the statutory responsibility to survey, document, and preserve Papua New Guinea's biological diversity.

Office of Environment and Conservation
P. O. Box 6601 Boroko, N.C.D.
Papua New Guinea

Bernice P. Bishop Museum (BPBM)

The BPBM, often called "the Museum of the Pacific," today serves as the State Museum of Natural and Cultural History for Hawaii. The museum supports programs in zoology, botany, anthropology, and education, all of which focus on the Pacific realm, its natural history, and its cultures. BPBM also serves as the Hawaiian Biological Survey.

Bernice Pauahi Bishop Museum
1525 Bernice St., P. O. Box 19000
Honolulu, Hawaii 96817-2704 USA
(tel) 808 848-4194
(fax) 808 847-8252

Papua New Guinea National Museum and Art Gallery (PNGNM)
The PNGNM is Papua New Guinea's premiere museum of culture and science, situated adjacent to the National Parliament building in Waigani, a suburb of Port Moresby. The PNGNM supports programs in natural history, archaeology, and culture.

Papua New Guinea National Museum and Art Gallery
P. O. Box 5560
Boroko, N.C.D., Papua New Guinea
(tel) 675-325-2458
(fax) 675-325-1779
PNGmuseum@global.net.pg

National Museum of Natural History (NMNH), USA
The NMNH is the Smithsonian Institution's museum of natural history and human cultures. It is among the world's largest natural history institutions, and it fosters field research and museum scholarship in fields as diverse as paleobotany, physical anthropology, and systematic entomology. The National Museum of Natural History is dedicated to understanding the natural world and our place in it.

National Museum of Natural History
Smithsonian Insitution
10th St. and Constitution, NW
Washington, DC 20560 USA

Acknowledgments

We first would like to thank the Government of Papua New Guinea for authorizing the RAP expedition to New Ireland, and for actively encouraging a collaboration between research scientists in Papua New Guinea (PNG) and those from overseas.

Second, we wish to thank the landowner groups from the Lak Electorate, who made our research team welcome on their traditional lands, and who assisted us in innumerable ways, helping to make the actual field exercise a success.

Bruce Jefferies and Rob McCallum, of the Conservation Resource Centre of DEC, provided remarkable logistical support for the expedition and ensured that we had every measure of comfort while we were afield. Michael Lucas Simu and Mathew Goga of Kanga Village, Oro Province, PNG, provided critical field technical support as local naturalists.

Kim Awbrey, Adrian Forsyth, and Cynthia Mackie (all formerly of CI headquarters), and Tim Werner provided important planning support.

Primary funding for this effort was provided by the U.S. Agency for International Development through the Biodiversity Support Program (BSP/WWF Agreement No. ME26, USAID #DHR 5554 A 00 8044 00). We are grateful to the late Molly Kux and Dr. Janis Alcorn for making this possible. Publication of this report was funded by a generous contribution from The Giuliani Family Foundation.

Report At A Glance

Southern New Ireland, Papua New Guinea: A Biodiversity Assessment

1. Dates of Studies

Weitin Base Camp (240–260 m):
14–24 January, 8–14 February, 1994

Lake Camp, Hans Meyer Range (1180–1200 m):
24 January–1 February, 1994

Top (High) Camp, Hans Meyer Range (1800–1830 m):
1–7 February, 1994

Riverside Camp, Weitin River (150–200 m):
3–16 February, 1994

2. Description of Location

New Ireland, 250 km long and 50 km at its widest, is the third largest island in Papua New Guinea (PNG) and is second only to New Britain in the Bismarck Archipelago. In this report, Southern New Ireland is considered to be the mountainous "bulb" that lies south of 4°S Latitude. It is dominated by the Hans Meyer Range to the north and the Verron Range to the south, the two being abruptly separated by the remarkable Weitin-Kamdaru fault that strikes southeast-northwest. This fault defines the flow of New Ireland's two largest rivers, the 36 km long Kamdaru and the 33 km long Weitin. Drainages are short, steep, and typically unstable, producing deeply cut valleys. The region has virtually no significant coastal plain; almost the entire landmass of the south is rugged hills and mountains, in most places extending down to within a kilometer of the coast. Southern New Ireland's 4000 km^2 of land supports fewer than 5000 people of which most live within a kilometer of the coast.

3. Reason for Expedition

In 1992, a Conservation Needs Assessment workshop identified southern New Ireland as one of twenty-five terrestrial areas of "very high importance" for biodiversity conservation in Papua New Guinea (PNG). This mountainous southern zone was also designated as a major "scientific unknown" for PNG. In 1994, Conservation International worked with the PNG Department of Environment and Conservation (DEC) to organize a rapid biodiversity assessment of the forests and wildlife of southern New Ireland. The objectives of the RAP expedition included (1) to assess the biodiversity of southern New Ireland, (2) to field-test rapid-survey methodology in PNG, and (3) to share expertise and methodologies with DEC staff scientists.

4. Major Results

The RAP team studied forest structure and surveyed the diversity of plants, butterflies and moths, reptiles, amphibians, birds, and mammals along an elevational transect from the coastline to near the summit of the Hans Meyer Range. The forests of southern New Ireland can be classified into five types, each distinct in floristics and structure: coastal lowland limestone forest, interior riverine lowland forest, hill forest, montane mossy forest, and mountaintop myrtaceous heath forest. The richest and most complex forests, found in hilly zones at elevations between 300 and 900 m, were impoverished relative to mainland New Guinea. Southern New Ireland has a high species diversity of moths, reptiles, amphibians, and bats. New Ireland's avifauna and non-volant mammal fauna are strikingly impoverished, possibly a product of both human activities and other past environmental events that also appear to have influenced the herpetofauna. Introduced predatory mammals (cats, dogs, pigs) are likely affecting the biodiversity of the area. Although southern New Ireland's flora and fauna are still incompletely inventoried, the Weitin Valley appears to conserve much of the island's biodiversity and is an important reservoir of the native flora and fauna used as resources by local people.

Number of species recorded

Plants:	500 tracheophyte species
Moths:	1088 species
Butterflies:	46 species
Amphibians (frogs):	7 species
Reptiles:	
Lizards	20 species
Snakes	12 species
Birds:	89 species
Mammals:	26 species

New species recorded

Plants:	6 species, including 2 orchids: *Saccoglossum* sp., *Dendrobium* sp. (Orchidaceae); *Corsia* sp. (Corsiaceae); *Freycinetia* sp. (Pandanaceae); *Psychotria* sp. (Rubiaceae)
Butterflies:	1 species of swallowtail, *Graphium kosii* sp., possibly one additional
Amphibians:	1 species of frog, *Platymantis browni*
Reptiles:	Possibly 2 species of lizard, *Sphenomorphus* sp. and *Carlia* sp.

New records for New Ireland

Reptiles:	1 snake: *Typhlops depressiceps*
Birds:	3 species: *Accipiter meyerianus* (Meyer's Goshawk), *Accipiter brachyurus* (New Britain Sparrow-hawk), *Podargus ocellatus* (Marbled Frogmouth)
Mammals:	1 flying fox, *Pteropus gilliardorum*, and 3 small bats (*Mosia nigrescens papuana, Nyctophilus microtis,* and *Philetor brachypterus*)

5. Conservation Recommendations

The forests of southern New Ireland are under severe threat from selective logging. Logging impacts the native biodiversity of the area in many ways, including removing nesting sites for Blyth's Hornbills, promoting the introduction of destructive invasive species such as feral dogs, cats, rats, and pigs into interior forests, and impacting the long-term ecological regeneration of those forests that have been culled.

Preservation of representative tracts of original (uncut) lowland forest in southern New Ireland should be an urgent priority. The uninhabited interior of southern New Ireland (south of 4°S Lat.) should remain under consideration as the "core" of a large forest conservation area. As part of the process of establishing a conservation area in southern New Ireland, a detailed land-use plan should be devised that balances future timber development with biodiversity conservation. Additional biological field investigation is merited in the highlands of the Verron Range and in the highlands northwest of the highest summit of the Hans Meyer Range. Focused searches should be made for vertebrate species inexplicably "missing" from New Ireland's current fauna.

Executive Summary

Introduction

Biologically, New Ireland has remained one of the least studied regions of Papua New Guinea (PNG), and the mountainous southern zone has been considered both a high priority for biodiversity conservation and a major "scientific unknown" (Beehler 1993). For these reasons, and because of the information needs of the now terminated conservation and development project established there by the United Nations Development Program (funded by the Global Environment Facility), Conservation International agreed to organize a rapid assessment (RAP) of the forests and wildlife of southern New Ireland. This was conducted in close collaboration with Papua New Guinea's Department of Environment and Conservation (DEC). The purpose of the RAP exercise was threefold: (1) to assess the biodiversity of southern New Ireland, (2) to field-test rapid-survey methodology in PNG, and (3) to share expertise and methodologies with DEC staff scientists.

The RAP Survey

A field team made up of thirteen Papua New Guinean and American scientists surveyed the forests and wildlife of southern New Ireland for 30 days (14 January–16 February 1994). Survey camps were established in the upper and middle Weitin River valley and also at two elevations in the Hans Meyer Range (1180 meters and 1830 meters above sea level), as follows (see Gazetteer for more details):

Weitin Base Camp (240–260 m):
14–24 January, 8–14 February, 1994

Lake Camp, Hans Meyer Range (1180–1200 m):
24 January–1 February, 1994

Top (High) Camp, Hans Meyer Range (1800–1830 m):
1–7 February, 1994

Riverside Camp, Weitin River (150–200 m):
3–16 February, 1994

Survey Results

The RAP team carried out surveys along a single elevational transect from the coastline to near the summit of the Hans Meyer Range, with some minor deviations from this plan, as time allowed. The plan also to ascend into the unsurveyed Verron Range to the south was prevented by logistical hindrances, and we strongly recommend that future researchers concentrate on that unknown block of montane forest. Below is a summary of our findings.

Forest Ecology

The structure and composition of the interior lowland forests of New Ireland are the product of short-term disturbance events, especially those related to river movement and local uplift. The upland forests are heavily mossed above 1500 m, and are locally stunted at or above 2000 m. The richest and most complex forests in southern New Ireland can be found in hilly zones at elevations between 300 and 900 m. Coastal zone forests are quite distinct from forests of the interior. This seems to be a product of the dominance of surface limestone in the coastal areas. Thus the forests of southern New Ireland can be classified into five types, a coastal lowland limestone forest, interior riverine lowland forest, hill forest, montane mossy forest, and mountaintop myrtaceous heath forest. Each is distinct in floristics and structure.

Botany

Floristic patterns are discussed from three 0.2 hectare transects established during the Rapid Assessment Program (RAP) biological survey of New Ireland's mountainous southern zone. The findings from transect-based inventory and *ad libitum* collecting indicate that humid forests in the surveyed tract are impoverished relative to mainland New Guinea. Species richness of canopy trees declines monotonically with elevation, but is also accompanied by increased percentages of Papuasian endemics within the montane communities. Nontree species exhibit similar patterns as overstory taxa, although the trends are less pronounced. Epiphytes are a particularly prominent component of the cloud-zone vegetation, accounting for more than one third of all taxonomic registers. Approximately 500 tracheophyte species were documented by the expedition, including a previously unknown orchid species in the endemic genus *Saccoglossum*, and five other novelties in the genera *Corsia* (Corsiaceae), *Dendrobium* (Orchidaceae), *Freycinetia* (Pandanaceae), and *Psychotria* (Rubiaceae).

Entomology

More than 10,000 moths were sampled at the four camps. At three camps, moth surveys were conducted in botanical survey plots in order to determine the relationship between plant and arthropod diversity. Moth species richness was relatively high, but not as high as comparable sites on the mainland. There is evidence of high endemism—numbers of species apparently peculiar to New Ireland. A general survey of butterflies and beetles proved productive, but the endemic population of *Papilio moerneri* was not found. The birdwing *Ornithoptera priamus urvillianus* was widespread and particularly common in the limestone coastal forest. A new endemic species of swallowtail butterfly, *Graphium kosii,* was particularly common above 2000 m.

Herpetology

During 32 days of field work at three main sites and a number of additional collecting localities in the Weitin Valley, southern New Ireland, we collected 351 specimens of amphibians and reptiles, representing 39 species. This total included seven species of frogs, 20 lizards, and 12 snakes, and comprises 81% of the 48 amphibian and reptile species now known to occur in New Ireland. One of the frogs is a new species now described as *Platymantis browni* and one or more of the lizards may also be undescribed. Three species are reported for the first time from New Ireland, including a significant range extension for *Typhlops depressiceps*, a blind snake previously known only from the mainland of New Guinea. Our study confirms that southern New Ireland is a center of high species diversity for the island.

Ornithology

Eighty-nine species of birds were recorded in southern New Ireland by Gibbs in 1993 and the RAP survey in 1994. Of these, 85 represent breeding land and freshwater species and four represent migrants and seabirds. Two species—Meyer's Goshawk (*Accipiter meyerianus*) and the New Britain Sparrowhawk (*Accipiter brachyurus*)—were recorded on New Ireland for the first time on this survey. The Marbled Frogmouth (*Podargus ocellatus*) was also added to the list by David Gibbs on a separate field trip.

Perhaps the most notable finding for the ornithological team was that which was absent from the survey. Extensive searches for a *Zoothera* ground-thrush, a *Cichlornis* thicket-warbler, and a medium-sized montane honeyeater failed to produce any evidence of their existence on New Ireland. Island representatives of all three of these are known to inhabit Bougainville and New Britain. Their absence from New Ireland is anomalous. Either these birds never inhabited the island or else became extinct. Further analysis of this phenomenon for other plant and animal taxa may prove useful.

A series of species of birds heretofore not known from New Ireland has been recorded from subfossil cave deposits by Steadman (1995). This includes several undescribed ground-dwelling rails and their relatives—one a large, flightless swamphen. We encountered none of these in our survey and postulate that they are now extinct.

Mammalogy

Twenty-six (26) species of mammals were recorded (three murine rodents, three marsupials, 17 bats, and three introduced/human commensals). It is thus evident that the New Ireland mammal fauna is impoverished, especially when compared to mainland PNG. The fruit-bats (Pteropodidae) are the most conspicuous components of the local fauna (with 12 species), although the less conspicuous insect-eating bats (Microchiroptera) are more species-rich, with some 20 species from the island. We found almost all of the species of fruit-bats known from New Ireland including *Pteropus gilliardorum,* a new record for the island and the only record since the unique holotype was collected on New Britain in 1959. In addition, we found three Microchiroptera new for New Ireland.

Testing Methodologies

This RAP survey differed from previous field expeditions in several respects, the main differences being the types of data collected and the focus of the work. Instead of stressing the simple accumulation of biological voucher specimens for museum and herbarium, the main goal was to collect data on the composition and structure of natural communities—data that are based on field sampling methods. Birds were censused repeatedly using vocal identification along a walked strip. Plants were censused along fixed transects, using repeatable ecological methods. Moths were censused using a 500-specimen repeated census. These types of methods produce relative abundance data and provide results that are more germane to conservation planning and ecological analysis. In general, the methodologies employed seem appropriate for PNG projects. Additional field projects will make it possible to test additional techniques for other taxa.

Training

Because of the great amount of work during the RAP survey, a general impression was that the training component received less attention than was deserved. But because of the shared duties of PNG and American counterparts, technology transfer and in-the-field training were inevitable in the instances when one of the counterparts was unfamiliar with a technique or specific group of plants or animals. There was considerable learning by both the American and PNG researchers. Still, it is believed that future field projects should define, in advance, the prime goal (either training or data collection). In other words, there should be two types of survey trips in Papua New Guinea—one stressing training, the other stressing biodiversity assessment.

Conservation Opportunities

Habitat Conversion and Logging

The forests of New Ireland are under threat from the actions of timber companies working in various logging concessions. Recently, logging activity in the Weitin Valley has been halted, and the company that was working there is shifting operations southward to the Jau River valley. The commercial timber in the forests of the Kamdaru River valley (just NW of the Weitin) have been harvested. It is apparent that most of the economically important lowland forest in southern New Ireland has been selectively logged.

What is the biological significance of this commercial logging on New Ireland? We surveyed in both logged and unlogged lowland forest in the Weitin valley. Selective logging produces gross physical alteration of forest that is starkly obvious at the ground level even if not so striking when viewed from an overflying aircraft. Nonetheless, the effects of this harvesting regime on biodiversity are not immediately obvious, and yet there was evidence that the long-term damage may be considerable.

Here are some of the concerns:

1. Nesting places for Blyth's Hornbills are apparently much reduced in logged sites. This hornbill, New Ireland's largest forest-dwelling bird, requires large natural cavities in tall trees for nesting. When the largest trees on a tract are harvested, the availability of natural cavities is reduced drastically.

2. Introduction of destructive human commensal animals into interior forests is promoted by logging activities. We found evidence of cats, feral dogs, Polynesian Rats, and Cane Toads in the Weitin valley. These non-native animals have a negative impact on the native fauna. The roads and clearings produced by logging foster the colonization of forest-edge habitats by these non-native species. The RAP mammalogy team found rat populations in the logged forest to be considerably higher that those in unlogged forest. It has been shown that on Pacific islands such as the Solomons, the introduction of non-native animals (especially rats and cats) is closely correlated with recent extinctions of native mammals.

3. The direct effects of selective removal of favored timber species is unknown, but may have an impact on the long-term ecological regeneration of those forests that have been culled.

Conservation Recommendations

We can offer the following preliminary recommendations, based on our observations in January and February 1994:

1. **Preserve remnants of Lowland Forest.** Preservation of representative tracts of original (uncut) lowland forest in southern New Ireland should be an urgent priority.

2. **Delineation of a Conservation Area.** The uninhabited interior of southern New Ireland (south of 4°S Lat.) should remain under consideration as the "core" of a large forest conservation area.

3. **Land-Use Planning.** As part of the process of establishing a conservation area in southern New Ireland, a detailed land-use plan should be devised that balances development with biodiversity conservation. The failure of the UNDP-sponsored integrated conservation-and-development project is, without doubt, a setback to this goal, but there is still an opportunity, should the government or local landowners take the initiative. As more and more landowner groups become engaged in management of their traditional lands, more and more of the undeveloped interior forest land should be subsumed under a detailed GIS analysis, with future utilization to be determined by landowner forums and a system of priority-setting that balances conservation versus sustainable development.

4. **Further Biological Study.** Additional biological field investigation is merited in the highlands of the Verron Range and in the highlands northwest of the highest summit of the Hans Meyer Range. Additional focused searches should be made for vertebrate species inexplicably "missing" from New Ireland's current fauna (especially those bird species expected for biogeographic reasons as well as those documented from the recent paleofauna).

Literature Cited

Beehler, B. M. (ed.) 1993. A Biodiversity Analysis for Papua New Guinea. Conservation Needs Assessment Vol. 2. Biodiversity Support Program. Washington, DC, USA.

Steadman, D. 1995. Prehistoric extinctions of Pacific Island birds: Biodiversity meets zooarcheology. Science. 267: 1123–1131.

Map 1. Location of New Ireland, Papua New Guinea

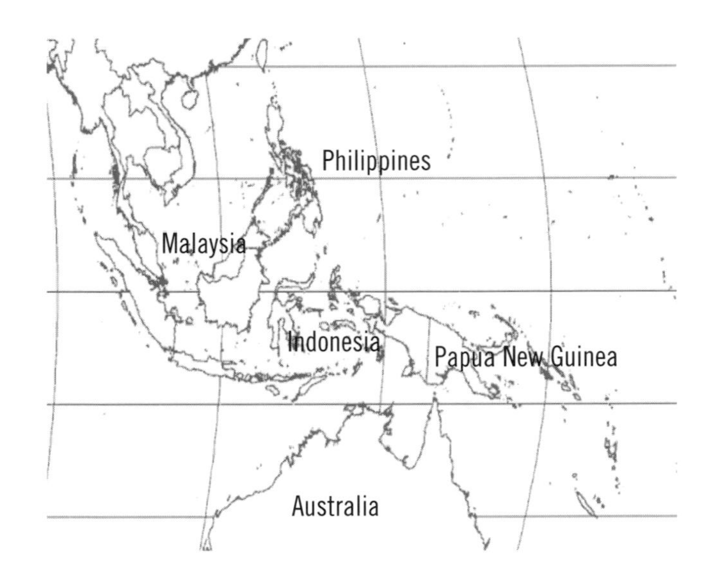

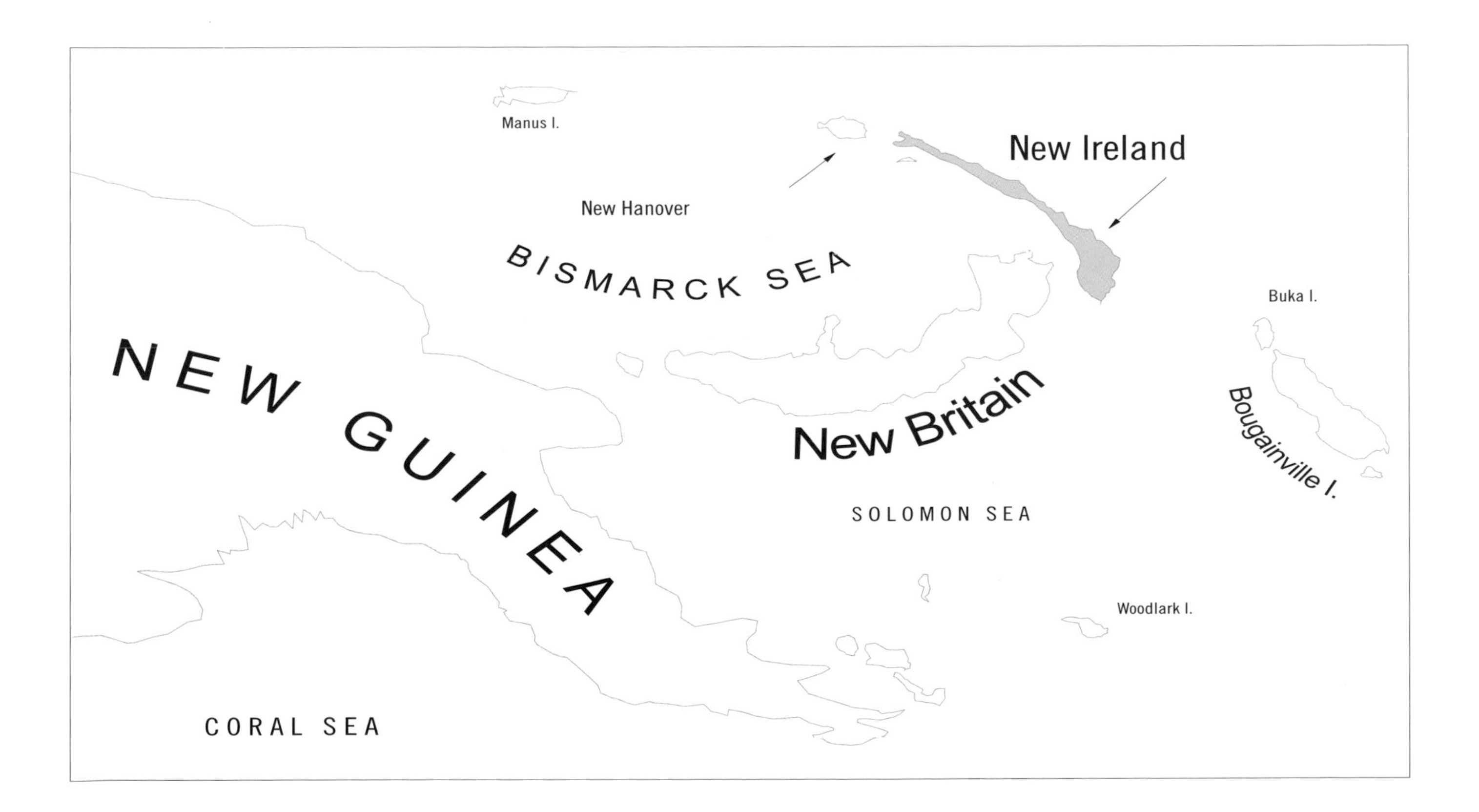

Map 2. Location of the four RAP survey sites (Top Camp, Lake Camp, Weitlin Base Camp, and Riverside Camp) in Southern New Ireland, Papua New Guinea

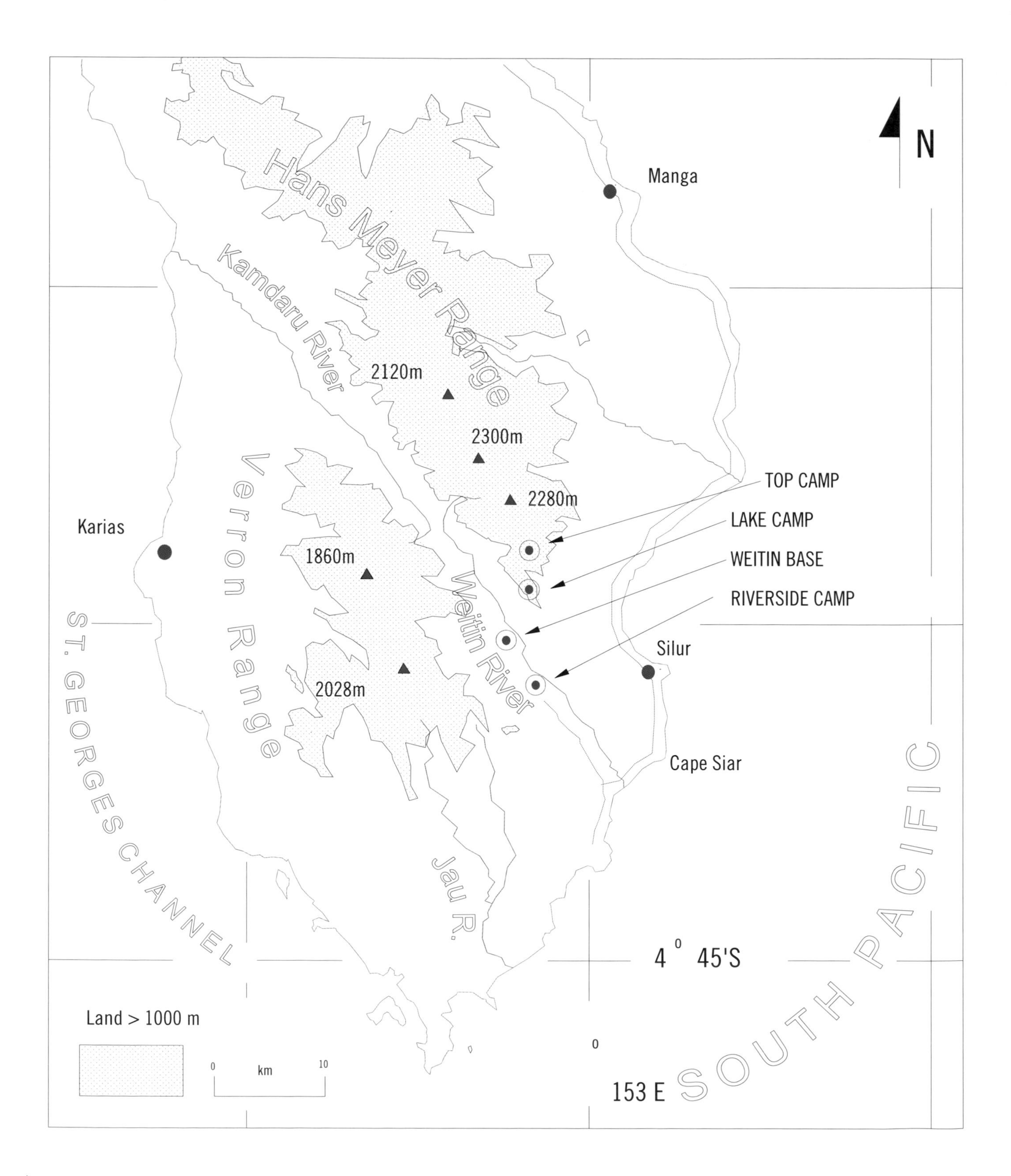

LANDSAT image of the area of the 1994 RAP survey in Southern New Ireland. Thematic Mapper (4,5,6).

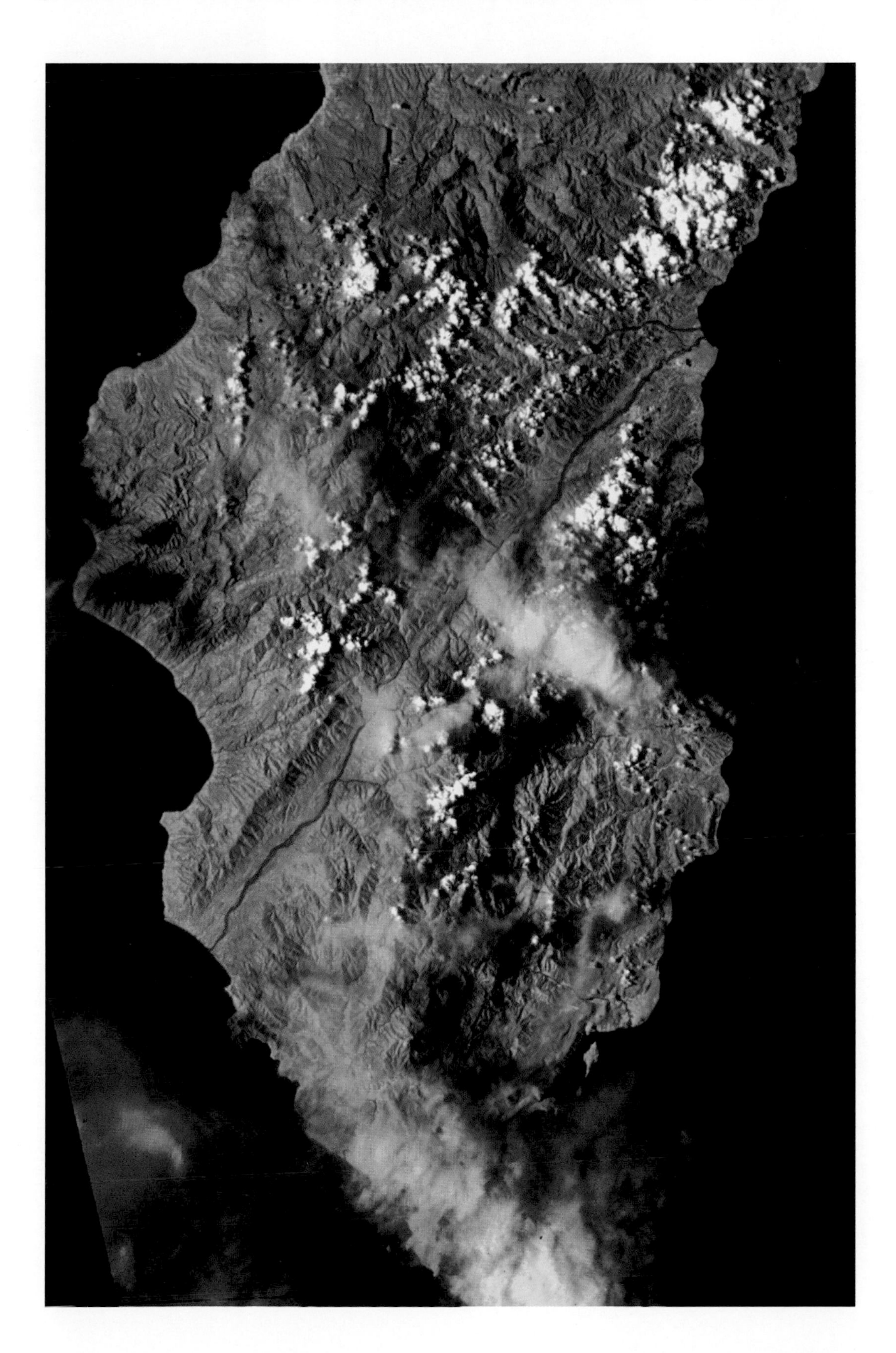

Bruce Beehler

High-elevation mossy forest at Top Camp (1800 m).

Bruce Beehler

Selectively-logged lowland forest near Riverside camp (250 m).

Bruce Beehler

The 1994 New Ireland RAP Team.

Bruce Beehler

Wayne Takeuchi, Louise Emmons and Allen Allison at Top Camp (1800 m).

Bruce Beehler

Tommy Kosi, Joseph Wiakabu, Robin Foster, and Wayne Takeuchi prepare botanical specimens at Lake Camp (1200 m).

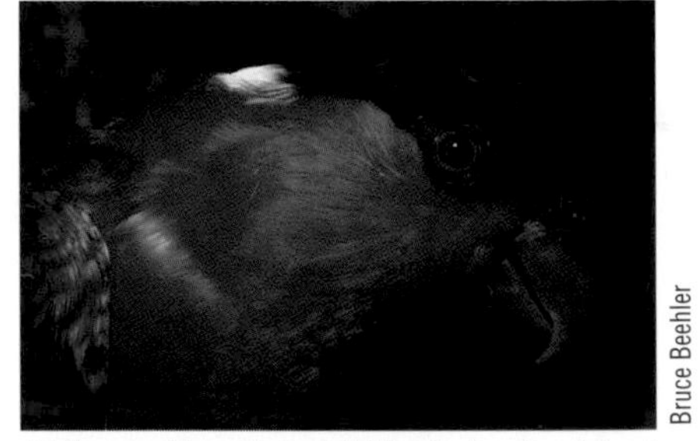

Bruce Beehler

White-naped Lory (*Lorius albidinuchus*), a New Ireland endemic, was fairly common singly or in pairs at Lake Camp (1200 m) but was not observed at the other RAP camps.

Allen Allison

Bothrochilus boa. This python, which occurs only in the Bismarck Archipelago and on Nissan, a small island between New Ireland and Bougainville, is thought to prey largely on other species of snakes. It is found in a wide variety of habitats and is particularly common in riparian areas.

Bruce Beehler

Monarcha verticalis, a Bismarck endemic, was recorded at the Riverside, Weitin and Lake Camps.

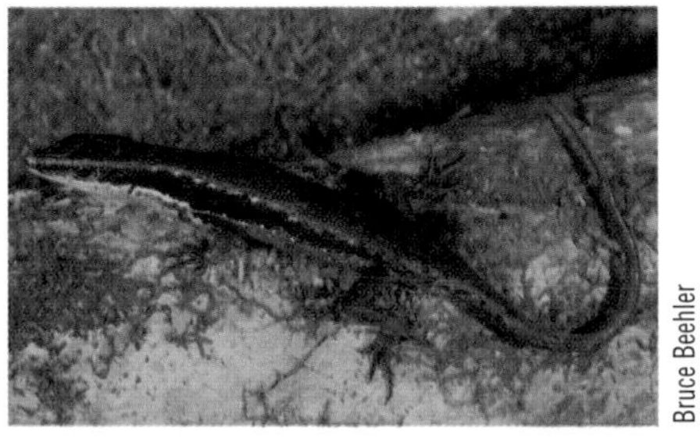

Bruce Beehler

Emoia bismarckensis. This skink, which occurs in the forest interior, is uncommon to rare. It is endemic to the Bismarck Archipelago.

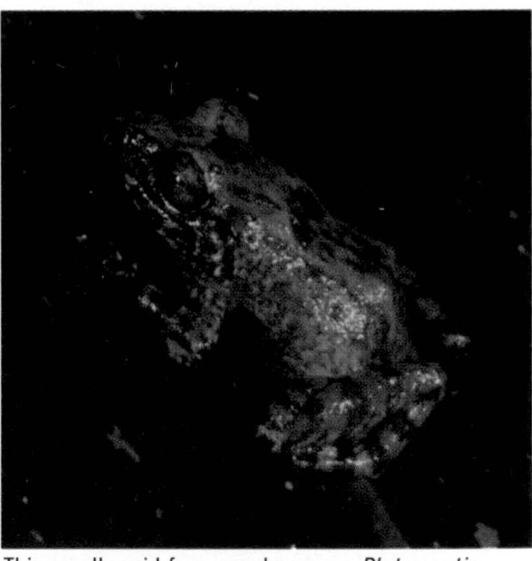

Allen Allison

This small ranid frog, now known as *Platymantis browni*, was discovered as a new species by members of the New Ireland RAP team. *Platymantis browni* occurs throughout the Weitin Valley and is probably the most abundant frog in the area.

Introduction to New Ireland and the RAP Expedition

Bruce M. Beehler and Kerry McGregor

In the following pages we set the scene with a brief review of New Ireland's natural and cultural history, followed by a summary narrative of the RAP expedition. This introduction is followed by six technical chapters that report the results of the RAP team's studies of the vegetation, botany, entomology, herpetology, ornithology, and mammalogy. Each of these technical chapters includes a bibliography and its own conservation assessment. As needed, data from the studies are reproduced in appendices that follow the technical chapters.

Introduction

Geography, Culture, History

New Ireland is the third largest island in Papua New Guinea (PNG) and second only to New Britain in the Bismarck Archipelago (Map 1). Some 250 km long and 50 km at its widest, and lying between the Equator and 5°S, New Ireland is part of an island arc that trends northwestward, including New Hanover and the Admiralty Islands. In the southeast, St. George's Channel separates New Ireland from New Britain's Gazelle Peninsula to the west by a mere 30 km, with the low, intervening Duke of York Islands as stepping stones between the two. It is fringed on the northeast by a parallel series of small islands that include Mussau, Tabar, Lihir, Tanga, Feni, and Nissan islands and their outlyers. This parallel island arc leads southward to the northernmost islands of Buka and Bougainville, part of PNG's Northern Solomons Province. The distance between New Ireland's southern coast and Buka is 175 km.

New Ireland is a high and complex oceanic island lying near the nexus of one of the earth's most tectonically active and geologically complex zones—the meeting place of the Australasian, Bismarck, and Solomons plates. New Ireland's mountains rise to their highest points in the far south, in several places to more than 2200 m, and the majority of its land area is swathed in tropical humid forest. Its current pattern of settlement favors the coastal zone and the northern half of the island. Thus the South is largely unpopulated and heavily forested.

New Ireland supports a population of ca. 60,000, of more than twenty-five language groups. Archeological evidence suggests that humans have inhabited New Ireland for 30,000 years (Robertson 1986). Western seafarers first sighted the island in 1616, when the Dutch explorers Schouten and le Maire skirted the eastern coast as they made their way northwestward to the Admiralty Islands (Reynolds 1972). It was Carteret in 1767 who demonstrated that New Ireland was an insular mass distinct from New Guinea, and he named it Nova Hibernia. By the 1840s Gower's Harbor, near the southern tip of the island, was a popular stopping and resting point for traders and whaling ships (Reynolds 1972).

By the late 1880s, the labor recruiters known as "blackbirders" began to appear in the region in search of cheap labor for the plantations in Queensland and the more developed Pacific Islands. In 1884 New Ireland was claimed by Germany as part of the colonial partitioning of New Guinea, and the first successful settlement attempts were accomplished by initiatives sponsored by the German Imperial administration, mainly in the north near Kavieng Harbor, in what was then termed Neu Mecklenburg. By 1903 thirteen plantations had been established. Methodist and Catholic missions had become well-established by the first decade of the 1900s and provided much of the local schooling and health services. Neu Mecklenburg was occupied by the Australians in 1914, and became part of the Mandated Territory declared by the League of Nations in 1920 and administered by Australia.

By 1940 a mere 200 Europeans and 450 Chinese shared the island with the local Melanesian inhabitants. The Japanese attacked and occupied the island in January 1941, and occupied it until surrender in September 1945. Post-war New Ireland saw little in the way of development or exploration, and even PNG's independence in 1975 from Australia did not produce any rush of modernization. But because of the coastal disposition of the population, and the long contact with western influences, New Ireland is home to a relatively sophisticated and well-educated people. Witness that a New Irelander, Sir Julius Chan, has served as PNG's Prime Minister in two separate tenures.

Southern New Ireland

Southern New Ireland is here considered to be the mountainous "bulb" that lies south of 4°S Latitude (Map 2). It is dominated by the Hans Meyer Range to the north and the Verron Range to the south, the two being abruptly separated by the remarkable Weitin-Kamdaru fault that strikes southeast-northwest. The region has virtually no significant coastal plain and a single natural harbor (Gower's Harbour, opposite Lambom Island). Virtually every settlement lies within a kilometer of the coast, apparently a phenomenon strongly influenced by past colonial resettlement activities.

Except for the floodplains created by the Kamdaru, Weitin, Jau and Jaulu rivers, almost the entire landmass of the south is rugged hills and mountains, in most places extending down to within a kilometer of the coast. Fringing barrier reefs occur, in varying widths, throughout the length of the coastline. A few coastal coconut plantations are scattered about—mostly in the west but also in the northeast (of Southern New Ireland).

The mountains are mainly composed of submarine Tertiary volcanics interbedded and in some instances capped with limestones and siltstones. Unlike New Britain, however, New Ireland lacks modern volcanic activity. Nonetheless a string of quaternary volcanoes, including Lihir, Feni, and Tanga, lie off the northeast coast. Also igneous intrusives occur locally—granite, granodiorite, and quartz diorite. The Weitin fault is the major structural feature in the region. There is evidence that the block northeast of the fault has risen considerably and shifted northward (Reynolds 1972). Southern New Ireland is seismically active because of its proximity to the deep seismic zone that lies to the east towards Bougainville.

The Weitin fault defines the flow of New Ireland's two largest rivers, the 36 km long Kamdaru and the 33 km long Weitin. Drainages are short, steep, and typically unstable, producing deeply cut valleys. The Weitin itself has produced a spectacular open and stony floodway for much of its length, and the casuarina woodlands fringing each bank are evidence that disturbance is a common feature of this environment.

There is evidence of more than 30,000 years of human occupation on New Ireland (Robertson 1986). Southern New Ireland, in spite of its current sparse settlement, supports considerable cultural diversity, with six Austronesian languages present. In general these groups show their main cultural affinities to those of the Duke of York Islands and of the Gazelle Peninsula of New Britain—not to those of northern New Ireland.

Southern New Ireland's 4000 km^2 of land supports fewer than 5000 people, slightly more than one person per square kilometer. Most live on the coast and few travel to the interior, even though there apparently is knowledge of traditional land tenure rights throughout. Coastal settlements tend to be small (10–20 houses) and subsistence gardens for these are usually within a kilometer of the coast.

Published knowledge on the region's flora and fauna is sparse, and is summarized in the individual technical chapters.

Setting the Scene for a Rapid Assessment

In 1992 the Conservation Needs Assessment workshop, held at the Christensen Research Institute, Madang, created a biodiversity priorities map for Papua New Guinea (Beehler 1993). Included among the twenty-five terrestrial areas of "very high importance" for biodiversity conservation was the mountainous expanse of southern New Ireland. The same workshop named this region in the list of major scientific unknowns (Beehler 1993). In part due to these priority designations, the Department of Environment and Conservation's Conservation Resource Centre selected southern New Ireland as the site of the first experimental integrated conservation-and-development (ICDP) project to be funded by the Global Environment Facility through the United Nations Development Program. This project has since ended; for more information about the project see McCallum and Sekhran (1997).

At the time of the RAP survey, localized and relatively restricted industrial-scale selective logging had been carried out in the Weitin watershed and its environs. Logging operations were focused on the few relatively flat tracts of forests in a largely mountainous environment. Some 100,000 cubic meters of timber had been exported from this area between 1991 and the time of the survey (McCallum and Sekhran 1997). As a result, much of the prime lowland forests accessible to the RAP program had already been selectively logged.

The Department of Environment and Conservation (DEC) in early 1993 invited Conservation International (CI) to organize and carry out a rapid assessment of the biodiversity resources of this site to help meet the goals of the ICDP project. This assistance included providing recommendations on the importance of the New Ireland area for global and regional conservation and recommendation on ways to mitigate threates in the area. The United States Agency for International Development (USAID) agreed to fund this rapid assessment through the Biodiversity Support Program (BSP).

Over a period of slightly more than six months, CI developed a field survey plan, vetted the plan with collaborating scientists in Papua New Guinea (PNG) at a meeting held in Waigani in June 1993, and then began the process of bringing the plan to fruition. This involved selecting an international team of field scientists to carry out the survey in collaboration with the DEC-selected PNG team; assembling the camping and field research supplies; completing a preliminary field itinerary; planning the support staffing and food requirements for the expedition; and making arrangements for visas, freighting, and plane and helicopter charters. These latter chores were facilitated in large part by the staff of the Conservation Resource Centre in Port Moresby.

In early January the team assembled in Port Moresby, made their last minute purchases, and then flew northeast to Rabaul (New Britain), and then on to the Silur landing ground on the eastern coast of southern New Ireland to begin the mission. The team was met there by Mike Hedemark with supplies, and thanks to his tireless advance work in New Ireland, within a day was ensconced at the Weitin Base Camp to begin the field survey of southern New Ireland—Conservation International's first RAP expedition to the Old World.

Narrative of the Expedition

Mike Hedemark and two assistants proceeded two days in advance to New Ireland to establish the Base Camp on the west bank of the Weitin River, ca. 17 km inland from the mouth (see Itinerary and Map 2). The team traveled by commercial jetliner from Port Moresby to Rabaul on 13 January 1994, and the next morning the main body of the team flew to Silur in three stages by small twin-engine aircraft. By 15 January the team was established at the Weitin Base Camp and had begun the survey.

Weitin Base Camp, 250 m, 14–24 January, 8–14 February 1994

This site, originally selected from examination of the 1:100,000 topographic map of the area, was visited and checked by a DEC team in October 1993. Situated at the confluence of the Weitin River and a major stream draining the east face of the Verron Range, by most measures it was an ideal site for a base camp, and offered ready access to water for drinking and bathing, and to various habitats for survey: open riverine regenerating vegetation, casuarina scrub, casuarina woodland, mixed-growth alluvial forest, and basal hill forest. The area, however, was dominated by the great open, sandy and stony floodways.

It so happens that the team arrived at the Base Camp during a severe drought, and the movement of warm dry wind up the valleys in conjunction with the sun beating unrelentingly on the rocky floodways produced a desiccating effect on the forest that appeared to drive much of the fauna to flight or into hiding. The area was found to be very quiet, day and night, and the zoological fieldwork, in particular, suffered.

Beehler arrived by helicopter from Rabaul on the early morning of 17 January, and found that the team was quite dissatisfied with survey conditions. After discussions that night, it was resolved that two exploratory teams would leave the next morning in search of upland camps where conditions would be more humid and more interesting for survey. Hedemark and Lucas led a team southwestward up a ridge of the Verron Range, whereas Antiko and Beehler led a team northwest up a ridge into the Hans Meyer Range. It was agreed both would spend a maximum of six days in search of good upland camps and then return and prepare for a move to an upland site deemed most promising.

The Hedemark/Lucas team was generally stymied by a near total lack of water at middle elevations in the Verron Range but did manage to clear a helipad for further exploration. Both teams were forced to subsist for several days on water obtained by cutting a species of upland bamboo. Ascending to a ridgetop at 1200 m, the Antiko/Beehler team located a small pond, and at 1800 m found a small spring. These ridgetop locations in the Hans Meyer Range were to become our main montane study areas—the Lake Camp and the Top Camp (see Itinerary and Gazetteer). An attempt by the Antiko/Beehler team to force its way to the summit of the Hans Meyer was foiled by a large area of tangled elfin scrub at about 2000 m on a false summit some 5 km SSE of the true summit of the range (2300 m).

The morning after the return of the Antiko/Beehler team a charter helicopter arrived (a day earlier than expected) and the Base Camp was hastily struck Over a 4.5 hour period the team, its equipment and all necessary supplies were shuttled by Bell 500 to the Lake Camp, where a small floating platform that the exploration team had constructed on the upland pond. This altogether inadequate structure had been constructed of treeferns and cut timber. The camp transfer was both difficult and hazardous for the pilot both because of excess overhanging vegetation surrounding the pond and the flimsy nature of the debarking platform. On the same day a parallel attempt to transfer a small exploratory party to the helipad in the Verron Range failed because of the inability of the pilot to put down in the restricted space as they were under-equipped to fell larger trees. The subsequent transfer to the Top Camp benefited from greater care and considerably more effort invested in clearing a landing space and constructing a steady landing pad.

Lake Camp, 1200 m, 24 January–1 February 1994

The party at the Lake Camp consisted of twenty, some twelve scientists and eight field assistants. Within a day of arrival, three shelters, a cook house, a men's house, and the "science center," were constructed of saplings and roofed with plastic. Mountain tents provided sleeping quarters for most of the scientific staff.

The Lake Camp environs proved to be the most pleasant of the habitats in which we worked. Although situated on a ridgetop, here the ridge had widened into a narrow and undulating plateau cloaked in a very beautiful forest dominated by a robust myrtaceous tree (*Syzygium* sp.) with shaggy exfoliating russet bark. Treeferns abounded in the understory, and the ground was covered by a thick cushion of leafy, loamy, and rooty duff that was both comfortable to sleep on and easy to navigate upon by foot. The entire team was pleased to have shifted from the hot dry "beach" of the Base Camp to the cool and verdant montane moss forest. The "lake"—really no more than a rather uninspiring shallow and leaf-clogged pond—offered a very congenial opening in the forest, and also attracted both breeding frogs and birds.

The fieldwork began in earnest at the Lake Camp. All disciplines worked full-bore here for seven days with considerable success. New Ireland's upland forests were what most of the fieldworkers had longed to survey, so there was very little complaining about conditions. There was little rain, but there was cloud, fog, and mist that produced local humidity, so the regional drought had a less significant impact at this elevation.

Arrangements had been made for a helicopter to shuttle half of the team to the Top Camp and the other half to the Base Camp on the Weitin for further lowland survey. In the early morning of 1 February the camp was broken in anticipation of the helicopter, which failed to appear. Part of the "Top" team departed on foot laden with heavy packs. The rest shifted by foot over the next day, and the day after that the helicopter arrived at the Top Camp with sparse quantities of supplies and food. Allison, having found no herps at this altitude after two nights and days of effort, descended to the Weitin Base Camp in the helicopter. While the Top Camp team carried out surveys there, the lowland team shifted camp from Weitin Base to the Riverside Camp, some 7 km downstream, situated on a Weitin tributary on the southwestern side of the main flow (Map 2).

Top Camp, 1800 m, 1–7 February 1994

Situated on a flat bench on the main ridge leading to the summit, the Top Camp allowed the team to work in the midst of New Ireland's high-elevation mossy forest—perpetually wet, and often cloud-enshrouded, with the weather changing several times a day, often rapidly. The helipad was set at the edge of a sharp drop-off that provided a spectacular view of the South Pacific—the coast a mere 11 km from this overview. We received at least some rain on all but one day at this camp. With some effort, this camp gave access to a range of elevations and vegetation types.

Near the camp the forest was well-developed but peculiar, with dominance by a single species of canopy tree, *Metrosideros salomonensis,* most of which were 40–60 cm in diameter, but with bowed and irregular trunks, all of which were heavily laden with a variety of epiphytic vegetation. The understory was fairly open here, with an abundance of treeferns in the poorly drained sites. The ground was spongy, wet, and irregular, often with subterranean cavities that posed difficulties for explorers. The only water available was from tiny ridgetop depressions or from rain captured off the plastic roofing.

Forty minute's climb up the ridge led to a false summit at 2000 m, where the forest was low, open, and dominated by the tangle of understory vegetation that prospered in absence of a dominating canopy. Interesting botanically, this tangle proved virtually impenetrable. A *Rhododendron* that was producing large and fragrant white blossoms was common here, as was a small red-flowered *Vaccinium. Metrosideros* remained the dominant tree even here.

The Top Camp was rustic in the extreme, in part because of limited flat ground, but also because of limited roofing plastic. A single work/dining shelter was constructed, with a tiny workmen's sleeping shelter adjacent. At peak occupation the Top team consisted of seven scientists and six assistants, but this was reduced to eleven, total, after the helicopter sorties. Hindered but not halted by the rain, mist, and cold temperatures, the team managed to carry out a full survey regimen during its tenure at the Top Camp. Low temperature recorded was 12.5°C. Because of the rapid envelopment of cloud and mist in the mid- to late morning, there was considerable concern about the ability of the helicopter to extricate the team and transfer it to the Riverside Camp. We were lucky to have a determined and skilled pilot with a Hughes Long Ranger, who moved the entire party in four sorties, the last two with ever more cloud obscuring the landing pad. He was much aided, apparently, by the judicious use of a global positioning device.

The entire team was briefly reunited at the comparatively luxurious Riverside Camp on 7 February (Daink Kuro and Tommy Kosi departed for Port Moresby on this day, hitching a ride on the helicopter on its return to Rabaul). The ten zoologists and seven assistants settled into Riverside Camp, to carry out their final segment of the field survey. The three botanical fieldworkers and several helpers hiked back up to the Weitin Base Camp to complete their plant and vegetation surveys there and make a short trip into the Verron Range, only to rejoin us after several days.

Riverside Camp, 165 m. 3–16 February 1994

The Riverside site was selected because it offered ready access to logged-over forest, and because it was accessible by truck. A gravel logging road followed the western side of the Weitin from its mouth northward to about two kilometers beyond

Riverside Camp. We could thus drive from Riverside to the Silur airstrip in about forty minutes (ca. 25 km). Our camp was about 0.3 km from the Weitin, and within a few hundred meters of the first rise of the foothills of the Verron Range. The forest of river flats was luxuriant but poorly developed, and had been much disturbed. It grew on stony and gravelly soils, a product of the constant braiding by the Weitin. Behind camp, the foothill forest was taller and more luxuriant, and much of the canopy remained intact, although it appeared that the loggers made a lot of effort to remove all of the giant *Octomeles sumatrana* emergents—a widespread forest canopy tree that abounds as an early successional pioneer on recently disturbed river cuts.

Riverside Camp was hot, with a maximum temperature of 28.5°C. The drought seemed to break by the time the full team had settled into this camp, and we received afternoon showers virtually every day we were there, with the volume of rain increasing day by day; 63 mm of rain fell on 14 February.

While based at Riverside, the team carried out surveys in the surrounding habitats and also took the opportunity to visit coastal sites for brief examination. This was aided by the availability of a Toyota Land Cruiser with four-wheel drive.

On 16 February we broke camp and were shuttled to the Silur landing ground for transfer by Twin Otter charters to Rabaul and then on to Port Moresby by Air Niugini jet. The team participated in an expedition assessment at the Department of Environment and Conservation on 17 February, followed by a presentation of the preliminary scientific results held at the National Library on 18 February, which brought to a close the team's collaboration in PNG. Many of the international staff departed the country shortly thereafter.

General Methods

The rapid assessment of southern New Ireland, Papua New Guinea, consisted of a single field trip in January and February of 1994. The rather large field team of thirteen scientists plus varying numbers of technical staff and New Irelander assistants made work out of a single camp difficult, so at times the team divided into separate parties, each occupying a separate survey camp. Not all personnel worked at all camps (see Itinerary). For details of the work of each taxonomic specialty, refer to that chapter in the series of reports that follows.

In general the survey worked an elevational transect from lowlands to mountaintops, followed by a final period of survey in a logged-over lowland area in the Weitin valley. An attempt to obtain advance information for selection of highland campsites by a two-hour helicopter reconnaissance provided little useful information (because of cloud and the strong disorientation produced by low-level helicopter maneuvering), but the available 1:100,000 topographic maps (PNG National Mapping Bureau, Waigani) were sufficiently detailed for this purpose. In lieu of expensive helicopter reconnaissance, future advance planning work in mountainous Papua New Guinea would be better served by light aircraft overflights from varying altitudes. The available LANDSAT imagery provided limited value for planning, mainly because of cloud cover over key highland areas, and the limited access the team had to New Ireland's interior. This imagery is useful for analysis, over larger areas, of gross distributions of habitat types and disturbance by logging, but given the rather limited range of habitats in southern New Ireland (mainly closed forest, anthropogenic grasslands, logged areas, rocky stream beds), the investment in imagery was of limited use to this RAP survey.

Surveys in the field tended to follow taxon-specific methodologies (see following chapters), but as noted earlier, unlike for typical museum expeditions, attempts were made to assess species populations and habitat selection, with collections being a secondary goal of the effort. Voucher collections were made of all focal taxa (plants, arthropods, birds, mammals, herpetofauna), but these were complemented by transect counts (plants, mammals, birds), and trap-and-release of mist-netted birds and bats. Voucher collections were shared between Papua New Guinean institutions (Christensen Research Institute: arthropods; PNG National Herbarium: plants; PNG National Museum and Art Gallery: mammals, birds, herpetofauna) and United States institutions (Bernice P. Bishop Museum: herpetofauna; U.S. National Museum of Natural History: birds, mammals).

Each international scientist was teamed with a Papua New Guinean counterpart, both to share knowledge and experience, and also to provide first-hand field training of RAP methods to PNG researchers. This method has been further expanded and improved in a more recent RAP trip to the Lakekamu Basin, Papua New Guinea (October–December 1996). Establishing a working team of proficient biodiversity field survey scientists in Papua New Guinea is a high priority for Conservation International.

Literature Cited

Beehler, B.M. (ed.) 1993. A Biodiversity Analysis for Papua New Guinea. Conservation Needs Assessment Vol. 2. Biodiversity Support Program. Washington, DC, USA.

McCallum R., and N. Sekhran, 1997. Race for the Rainforest: Evaluation Lessons from the Integrated Conservation and Development "Experiment" in New Ireland, Papua New Guinea. Department of Environment and Conservation/United Nations Development Programme.

Reynolds, J.P. 1972. New Ireland District. Pp. 847–859 *In*: P. Ryan (ed.) Encyclopedia of Papua and New Guinea. Melbourne, University Press, Melbourne, Australia.

Robertson, N. 1986. Matenkupum, a Late Pleistocene cave on New Ireland, Papua New Guinea. Honours thesis, Department of Archaeology, La Trobe University, Australia.

Chapter 1

The Forest Vegetation of Southern New Ireland

Robin B. Foster

Abstract

Southern New Ireland is both mountainous and heavily forested, and receives sufficient rainfall throughout the year to support humid closed-canopy forest from the rocky limestone coasts to the highest summits. From overflights and examination of LANDSAT imagery, one can discern what appear to be five major types of forest in southern New Ireland—coastal limestone forest, lowland forest, hillslope forest, mid-montane cloud forest, and high mountain mossy forest. These are described in varying degrees of detail below, along with a number of peculiarities that mark the vegetation of particular locales.

Introduction

The botany and vegetation of New Ireland has been little studied (Paijmans 1976). Scattered and uncoordinated botanical collections were doubtless made on and off during the colonial years, but little is reported in the literature (see Johns 1993, Stevens 1989, Conn 1994). The single comprehensive expedition to the uplands of southern New Ireland was that of D. Sands, W. R. Barker, and J. Croft in 1975, although it is not annotated in any publication.

Methods

We sampled the forest vegetation in the Weitin Valley and at both the mid-elevation (Lake Camp) and high-elevation (Top Camp) camps (see Gazetteer and Itinerary). Sample tracts of vegetation were selected based on what *a priori* appeared to be distinct plant communities during overflights, initial exploration, and simply because sites were distinct in elevation, topography, or substrate and thus suspected to support different plant communities. The vegetation was sampled using a "variable transect" method developed by the author. The method is based on sampling a standard number of individuals rather than a standard area. The same number of individuals is sampled for different size-classes or life-forms and compared between areas. It does not require the establishment of plots or fixed-length transects, but the area sampled is measured approximately and densities can be calculated. The placement of the transects was determined mainly by convenience, and it was usually most convenient to have them parallel to the narrow exploratory trails cut by the expedition team. In the mountains these trails usually followed ridge crests; consequently there is a bias of sampling in favor of ridges rather than hill flanks or ravines.

In this study, 100 individuals were sampled when possible for each class at each location, with a few exceptions such as for trunk epiphytes, herbs, and some longer, canopy-tree transects. The width of the transect was changed to be appropriate for different density of stems: 20 m wide for canopy trees (stems > 30 cm diameter at 1.3 m above ground) and medium trees (10–29 cm); and 2 m wide for smaller stems and herbs. A width of 5 m was used for some of the early-successional transects on the floodplain. To insure that smaller stems were not all clustered at one end of a transect, only 10 were sampled for each segment of 10 canopy trees (the class of plants with lowest density). A single transect sample of 4 classes (large trees, medium trees, small trees or shrubs, and herbs)—400 plants—could usually be done in one day. One hundred canopy trees usually occupied an area somewhat larger than one hectare.

Trunk epiphytes were sampled at the mid-elevation (Lake Camp) site on each of 100 medium trees in the transect. Epiphyte species were recorded only once for each tree up to a height of 3 m on the host stem. At the high-elevation (Top

Camp) site, 30 medium trees were sampled for trunk epiphytes in groups of 10 at the ends and center of the 100 tree transect. At other sites there were so few trunk epiphytes that they were ignored.

This was the first time that variable transects were tested for comparison of such diverse vegetation conditions with less than a month of effort. The results provide a ready quantitative description of the composition and comparative species richness of a wide variety of communities in a very short time. A probable improvement that would cover more of the inherent variance in species abundance and distribution would be to do twice as many transects of only 50 individuals each, with the two transects in each "habitat" separated by about 500 m when possible.

Results

The following is a first attempt to describe the plant communities of Southern New Ireland. The distinctions are made on the basis of what seem to be significant differences in floristic composition or vegetation structure. As it is based only on the area of our exploration around the central-southern Weitin Valley and a few observations from planes and trucks while coming and going, it can only be considered tentative as a description for the whole region. What we may have found as clear distinctions in floristic composition may in other areas be very gradual transitions in plant populations.

Coastal Communities

The plant communities along the coast, including bits of mangrove, beaches, and rocky outcrops, were not investigated as part of this expedition but the species are well-described in the flora of Peekel (1984). Because of the roads and long history of human establishment along the coastline, this vegetation has been much disturbed, and the only place to study it in anything like intact form would be on the southernmost coastline—mostly rocky. Rimming the island on the lowest slopes, but not in the Weiten Valley, is a broad band of limestone rock. While not explored on foot on this trip, from the air it could be seen to be mostly dominated by *Vitex cofassus* (Verbenaceae), a deciduous tree relatively rare in the Weitin Valley. This vegetation is much more extensive on the neighboring island of New Britain which has large expanses of limestone, in the interior as well as along the coast.

Valley Floodplain and Talus Slopes

The floor of the Weitin Valley (ca. 100–180 m elevation) and the lower parts of its lateral tributaries have flat-bottomed, broad floodplains with very unstable braided streams shifting over open beds of alluvial rock and gravel. In the vicinity of Riverside Camp the erosion during the last few hundred years in the main valley has been stronger on the southwestern side. This has left a series of natural levees on the northeastern side, each succeedingly older than the previous in the direction of the valley wall. Pieces of vegetated levees of all ages remain as elongated islands throughout the floodplain.

The usual successional sequence is the following:

1. Herbaceous colonization of the bare gravel dominated by a wild cane (*Saccharum:* Poaceae), interspersed with dozens of other herbs and seedlings and small saplings of trees.

2. Short-tree stage (tree stems 1–10 cm diameter) dominated by 5 m-tall *Casuarina equisetifolia* (Casuarinaceae) (28% of stems in transect) and the small tree, *Timonius* sp. (Rubiaceae) (24%); and a ground layer dominated by a *Nephrolepis* sp. (Pteridophyta) mixed with 32 other species of herbs and juvenile woody plants, especially the shrub *Polyscias cumingiana* (Araliaceae).

3. Medium-tree stage in which the 10 m-tall *Casuarina* make up most of the trees over 10 cm diameter (90% of stems in the transect), along with 5 small tree species, and an understory with increasing numbers of saplings of large-tree species. *Nephrolepis* persists in the ground layer.

4. Large tree stage in which the 30 m-tall emergent *Casuarina* (48% of stems over 30 cm diameter) are mixed with 20 canopy species, mostly young adults of large-tree species, while the original small-tree species and *Nephrolepis* ferns have almost disappeared. *Polyscias* persists in the shrub layer.

5. Old floodplain forest of large flood-tolerant species, without *Casuarina*, has a 30 m tall heterogeneous canopy (31 species over 30 cm diameter), the most abundant being *Pometia pinnata* (Sapindaceae) (23% of stems in transect), and an understory of 38 medium-size (10-30 cm diameter) species, especially a *Dysoxylum* sp. (Meliaceae) (13%) and a *Myristica* sp. (Myristicaceae) (11%).

Giant Forest

On small alluvial fans and talus slopes at the base of the valley walls (about 200 m elevation) is a mix of species, apparently invading from the older floodplain below and from the lower hill forest above. This is the most impressive-looking forest of the region, with many giant trees up to or exceeding 50 m height, especially of *Octomeles sumatrana* (Datiscaceae). Although *Pometia* is the most abundant tree over 30 cm diameter (20% of stems in transect) it is only rarely a giant.

The next most abundant tree, *Dendrocnide cordata* (Urticaceae) (11%), is one of the giants. The medium-size trees have greater dominance (20% and 15% respectively) by the same *Myristica* and *Dysoxylum* as above, as well as abundant *Dendrocnide warburgii* (13%).

These stands sit on deep, well-drained, heterogeneous soils and are protected from the ravages of the river and sheltered from the wind. As this high-forest ages, the giant trees gradually disintegrate or fall and the forest erodes leaving impenetrable liana tangles with only occasional lone survivors of forest giants. The entire process probably takes a few hundred years, and it is not obvious whether these natural vine tangles spontaneously revert to forest, or stay indefinitely until some major disturbance such as landslide or river-cutting re-initiates a successional process.

On the southwest (Verron) side of the valley, long stretches of what seems to be young forest have no *Casuarina*, and sometimes support large patches of *Pandanus* (Pandanaceae).

Hill Forest

The forest on the slopes and ridges from ~250 m up to 800 m are the most heterogeneous from place to place. Each ridge has its own peculiar combination of dominant species. It is surmised that seed dispersal between ridges is much less frequent than dispersal within a ridge. During colonization and establishment of treefall-gaps and infrequent landslides there would result a perpetuation and enhancement of chance differences between the ridges.

This is also the richest in species at the local scale of all the plant communities sampled. For most of the hill forest the most abundant canopy tree (though not the largest) is *Pometia pinnata* (20% of stems in transect at 350 m elevation). No other species made up even 10% of the stems. Also among the medium-trees, few were common, with only *Neoscortechinia forbesii* (Euphorbiaceae) making up 13%. Among the small stems (1–10 cm diameter), not a single species made up more than 4% of the stems. The canopy height is variable, usually 30–40 m tall. There are few or no epiphytes or moss on the tree-trunks.

There is some difference between low hill forest and upper hill forest in species composition, and this may reflect a slightly cooler and moister habitat moving up slope. *Pometia* becomes less abundant. A few epiphytes and occasional moss begin to show up on the tree trunks above 500 m—much more apparent in ravines than on ridges. A transect at 750 m elevation had the greatest number of species of any transect sample, and no species made up even 10% of the large or medium-size trees. This diversity is apparently in large part due to the presence of straying tree species from the mid-elevation cloud forest above. Comparing woody plants of different size classes, in the low hill forest transect the greatest number of species is in the smallest size class, whereas in the upper hill forest the greatest richness is in the large trees.

Between 800 and 900 m, at a point where the clouds usually reach their lower limit, is a much more abrupt transition to mid-elevation cloud forest. On the north side of the Weitin Valley the steepest slopes happen to coincide with the lower cloud limit. Thus although there is some transition before the steepest part, the frequent landslides keep a broad band in a continuous state of succession and tangle. This band largely separates the hill forest community from the mid-elevation cloud forest. On the southwest side of the valley the transition is more gradual.

Apparently for some geological reason (though perhaps by chance at this time) the steep slopes south of the Weitin valley are especially subject to landslide, at least much more so than on the north. Colonizing these slides are dense stands of spiny bamboo, tangles of *Rubus* (Rosaceae), and other spiny plants. Combined, in many spots they make it a great challenge to reach the open forest of the moderate slopes and ridgetops above them.

Mid-elevation Cloud Forest

The forest between 900 and 1500 meters is characterized in the dry season by sunny days and afternoon and evening envelopment by clouds and sometimes a light drizzle of rain. Consequently the plants are subject to some level of drought stress in the daytime during this period. In the wet season clouds are apparently accompanied by rain for much of the afternoon.

On more exposed ridges at this elevation and those further from the central mountain massif, the mid-elevation cloud forest is apparently subjected to greater risk of drought. Consequently the diversity of trees and epiphytes drops in these drier areas, though the species that remain are still characteristic of this zone. Moss is scattered throughout but does not obscure the recognizability of different species' trunks nor cover the ground. Terrestrial ferns are diverse and frequently form a thick cover. There is a deep organic layer on the flatter areas, and small trees and treeferns are easily toppled with a firm push.

A transect at 1200 m elevation in the wettest and flattest part of this forest has a canopy dominated by *Weinmannia* cf. *blumei* (Cunoniaceae) (19% of trees) and a *Pouteria* [=*Planchonella*] cf. *linggensis* (Sapotaceae) (11%). In similar terrain but at 900 m elevation southwest of the Weitin, the same *Pouteria* made up 12% of the large trees but *Weinmannia* was absent—though present at higher elevations. On a transect of wet but well-drained ridges at 1200 m, relative abundance of *Weinmannia* increased to 30% of canopy trees, *Pouteria* became uncommon, while *Gordonia amboinensis* (Theaceae) and *Calophyllum vexans* (Clusiaceae) became more important with 15 and 11% respectively of the large trees. As the same ridge extends away from the central core of mountains (at the same elevation) and the habitat becomes increasingly dry, *Gordonia* becomes the dominant with 36% of the large trees in a transect, *Weinmannia* drops to 16%, and *Platea excelsa* (Rhamnaceae) gains importance to make up 11%.

In the wetter areas, the dominant medium-size tree was *Mallotis echinatus* (Euphorbiaceae), the same species on both sides of the valley (14% north, 25% south in the transects). Tree-ferns are conspicuous in the wetter spots but are not especially common.

There is an extraordinary abundance and diversity of trunk epiphytes, mainly ferns and orchids. Below a height of 3 m on the trunk, a transect sample of 100 trees had a total of 59 species of epiphytes. The average tree trunk had 30 species of epiphytes, and the median epiphyte was on 10 trees. Needless to say, there were other epiphytes higher up on the trees.

High-elevation Mossy Forest

On the highest ridges from 1600 m up to 2300 m is a permanently wet zone with heavy moss cover on trunks as well as the ground. In the dry season, clouds move in by 1000 hrs and remain until the very end of the day when they dissipate, leaving a clear mountaintop above a lower cloud layer in the evening. In the wet season the clouds are apparently more accompanied by continuos rain.

The 10–20 m forest canopy is dominated by *Metrosideros salomonensis* (Myrtaceae) making up 75% of trees more than 30 cm diameter in the transect at 2000 m on Angil Mountain. (A species of *Metrosideros* is the dominant tree in many of the upland forests of the Hawaiian Islands.) Here they are large-trunked trees that are mostly bent over but with erect secondary stems reaching the canopy. Along the main stem are additional root systems supporting the secondary trunks. Covering all but the undersides of these massive "shrubs" is thick moss. The dominant medium-size tree is an *Ascarina* sp. (Chloranthaceae) (37%), and the dominant of the 1–10 cm diam. stems is an *Ilex* sp. (Aquifoliaceae) (38%). There is very little bamboo except for some thin-stemmed *Nastus*. The treeferns are mostly dwarf, about 1 m tall.

No juvenile *Metrosideros* were encountered, only sprouts of older trunks. Probably they get established on open landslides on the sides of the ridges and "walk" in to the flat ridgetops by falling over and sprouting along the trunk.

The deep moss may explain the low diversity and density of other trunk epiphytes at this altitude. It appears that to get established, many epiphytes need a firm attachment to the trunk from the beginning. Either that or the moss is in some other way inhibitory to vascular epiphyte colonization. Perhaps many epiphytes only get a foothold when a branch breaks or from animal activity. In a sample of 30 trees along the transect at 2000 m, there were 41 species of epiphytes below 3 m on the trunk. The average trunk had only 7 species of epiphytes, and the median epiphyte was on 3 trees.

On the uppermost slopes, as in most very wet, cool equatorial cloud/mossy/elfin forests, the stature of the trees is lower and the understory is an increasingly impenetrable maze of horizontal branches and roots covered with mounds of moss and frequently hiding deep holes down to the hidden terra firme. A few species such as a *Drimys* (s.l.) sp. (Winteraceae) were encountered only in this extreme habitat.

Mountain-top Thicket

On flat crests of the mountain tops on either side of the Weiten Valley are open thickets of vegetation and mossy mounds 3–4 m tall over hidden remnants of trees, with scattered small trees 5–7 m tall. These thickets (on the Hans Meyer side) were investigated with some difficulty to see if they contained a distinct plant community. They did not. Except for a few species of epiphytes and a few herbs not seen elsewhere, the flora of the thicket was a reduced and apparently degraded variant of the forest flora surrounding it. Unless humans were at some time responsible for clearing the area which is now slowly recovering, I would suggest that the cause of these thickets is localized fire caused by frequent lightning strikes. Some of the old trunks look charred and most of the smaller trees and shrubs are resprouts. The openings that result from the mortality of the larger trees has allowed a proliferation of shrubs (especially Myrtaceae), herbs, and hemiepiphytes such as *Rhododendron superbum* (Ericaceae).

Other Local Peculiarities in the Vegetation

Apart from the plant communities described above, there are other distinct forest communities that inhabit southern New Ireland. Already mentioned are the differences between different ridges in the hillslope forest. More dramatic are differences between the north and south sides of the Weitin Valley. In addition to having a more gradual transition between the communities because of the more gradual increase in elevation, there are significant differences in the flora even where the habitat appears to be the same. Several important tree species such as the red-trunked *Dillenia schlechteri* (Dilleniaceae) and an unidentified tree (Euphorbiaceae) were encountered only southwest of the Weitin. Most noticeable from an aerial view is the emergent deciduous giant *Serianthes minahassae* (Fabaceae-Mimosoideae), which occurs in local clusters over much of the upper hillslope only in the Verron Range. It remains to be seen if this differentiation is (a) only a local phenomenon—a more exaggerated version of the differences between adjacent minor ridges, (b) a reflection of a geological difference that is not easily noticeable to the eye, or (c) whether the Weiten Valley is actually an important barrier isolating the two halves of southern New Ireland. Lack of permanent water prevented the establishment of field camps southwest of the Weitin.

Comparing east and west on a larger scale rather than northeast and southwest on a more local scale, it is notable that there is a change in the prevailing wind direction over different parts of the year. When the eastern slope of the island is wettest, the western slope is driest, and vice versa. This could conceivably cause a differentiation in the flora of east and west slopes of New Ireland especially at the lower elevations which suffer more pronounced drought. One could imagine that immigrant seeds would find one side or the other more suitable for establishment depending on the season. Once established, they might still find it difficult to migrate between sides of the island because the timing of seed production might be ill-suited to establishment on the other side. Since we did not explore the northwestern end of the Weitin-Kamdaru rift, it will remain for others to investigate. In terms of protecting plant diversity in this region, it could be important.

Diversity Patterns

The variable transect data for canopy trees (> 30 cm diameter) in mature forest show that the hill forest has between 30 and 50 species for every 100 trees. Both floodplain and middle elevation cloud forest have between 20 and 35 species per hundred, and the high elevation mossy forest has 5–10 species per hundred. There are a very few species that overlap two or three of these communities.

The Weitin Valley floor, as it is in a state of perpetual succession, is not expected to be especially rich in plant species. Perhaps if the forest lasted longer before being carved away again by the river or being smothered by a landslide, there would be time for more species to accumulate. Presumably many more of the hill forest species could potentially grow on the older and rarely flooded parts of the valley floor. The number of species increases as the Casuarina forest ages and is finally replaced, but never quite reaches the richness of the hill forest.

In southern New Ireland the hill forest is the habitat richest for trees and shrubs. Hill forest reaches maximal diversity at the upper end—its ecotone with the mid-elevation cloud forest. Just the addition of species from higher-up into what for them is a marginal, drier environment below is enough to explain this increase. Thus it is not an optimal habitat for more species, but rather a habitat that is at least temporarily tolerated by species of two different communities.

Upward from 1000 m there is a progressive decrease in tree and shrub species, though with variation apparently according to how much the site is subject to drying conditions. Epiphytes however are richest in species (and abundance) from 1000 to 1600 m, then gradually decrease at higher elevations. At elevations below 900 m, epiphytes abruptly disappear.

Disturbance

The narrow coastal plain of New Ireland has been subject to the most sustained recent (post-contact) human disturbance. The fertile soils on the calcareous rock that fringes the island, the year-round availability of fresh water from the cloudy mountaintops to the coast, and the many marine resources available have all contributed to making the coast a much more desirable place to settle than the interior. There is little evidence of recent or past human activity in the interior beyond the recent logging. Trails are very few and infrequently used. Virtually none of our local assistants, though living not far away, had ever visited either the upper Weitin valley or the upland forests where we made our mountain camps. The abundance of feral pigs and wallabies suggest that hunting pressure in the interior is low.

Pigs

The human introduction of pigs may be significantly effecting the populations of many plant species. The more obvious effects—the tremendous trampling and disturbance underneath large fig trees and other favored food species—may be of minimal importance and may even improve dispersal of some species. With flying foxes available, terrestrial dispersers may never have had an important role on the island, and the destruction of a mass of larger seeds on the ground under the parent may be nearly irrelevant to its dispersal. But another example indicates more impact. The only two individuals of the large *Marattia* fern encountered were both in places that a pig snout could not reach. The stem and root of this fern is known to be consumed frequently by feral pigs. This fern was presumably much more abundant in the past, but how abundant we will never know, unless perhaps there are some large and long-term exclosure experiments that keep feral pigs out. It is expected that they have had the same effect on many other species. It appears well known that these pigs migrate for weeks at a time to different areas with seasonal concentrations of fruit crops. A study of what plants pigs disperse, what they destroy, and how efficient they are away from big fruit concentrations would be most useful.

Logging

Removal of timber is evident throughout the lower slopes nearest the coast. The logging operations have been very badly done. According to our informants, the loggers are not selective about species, but rather take almost anything with a large, intact, straight trunk. Observation of the previously logged forest confirms this. The logged areas are easy to recognize because the remaining vegetation becomes smothered with vines. This is disastrous for the future of these forests. Certainly the removal of most of the large trees rather than more selective cutting has much to do with it. Possibly the vines (just a few species are involved) are unusually aggressive. In any case, vine-coated vegetation like this tends to stay that way, possibly for hundreds of years. Trees cannot grow up under it and if they do happen to get started, the vines rapidly catch up with them. What this means is no forest resources for the future in the areas that have been logged.

The only option is to try various experiments to reduce the vine cover while promoting opportunities for tree growth: possibly by burning, shading, bulldozing, introducing specific pests or diseases, or even using mild herbicides if necessary. Most of these options are difficult or costly. These forests could have been managed well with careful selective extraction, or selective small-scale clear-cutting, and they could have provided an endless supply of resources and income for the people who live on New Ireland.

Opportunities for Conservation of the Native Vegetation

1. Southern New Ireland suffers little local population pressure or local demand for resources from the large, relatively pristine interior parts of the island. The local population has become relatively sophisticated through long and difficult experience about what level of exploitation of resources is important to them.

2. The Weitin Valley is a logical centerpiece for a core reserve in the mountainous central area of Southern New Ireland. It has great landscape beauty, it contains the range of natural communities from valley floor to the highest mossy forest peaks on the island, it probably contains the majority of the important plant diversity on the island, relatively easy access is possible, and the threats to it are relatively easy to manage.

3. The area that surrounds such a reserve or park is suitable as a buffer zone where well-managed forestry practices and possibly limited hunting could supply resources and a sustainable source of income for the local communities without seriously endangering the biodiversity of the area. The buffer zone is also suitable for experiments in forest restoration with native species on those areas already logged.

4. The hillslope forests have the greatest diversity of tree and shrub species and deserve special priority in protection. The middle and high-elevation mossy forests contain the highest diversity of epiphyte species and in addition are absolutely critical to keep intact to supply the fresh water streams as a year-round resource for the communities on the coastal lowlands. Without the mossy forest to filter water out of the foggy air, these streams will dry up during periods with no rain.

5. The remaining giant-tree forests at the base of the slopes are important to protect and are impressive as a potential tourist attraction. Similarly the two different cloud-forests are exciting for tourists, especially for the great density of epiphytic ferns and orchids and the enchanting fairyland aspect of the deep mossy forest. The latter might be more easily reached from the backside (i.e., the logging roads and trails northeast of) rather than from the Weitin Valley.

6. The valley itself and the area northeast of the Weitin seem most suitable for developing ecotourism in the reserve, especially because of the access roads that already exist, and because they cover the range of habitats from the highest to lowest in a short distance. The remaining areas of reserve could be limited more to scientific work and species protection. Experimental research or destructive sampling could be limited to the buffer zone.

7. There is a good opportunity to produce basic environmental information for the local people, especially about the hydrology and the forest of this area and how it can affect them, as well as more about the flora and fauna and why it is both interesting and important.

Recommendations for Additional Research

1. Explore the southern end or eastern side of the island to see if good examples of additional habitats, especially limestone forests and coastal plant communities, can reasonably be included either contiguously or separately with the core area studied here.

2. Conduct an inventory similar to this one of the forest composition on the western side of the island and more to the south.

3. Map the major habitats in the region.

4. Establish some permanent plots in each of the major habitats and monitor tree growth.

5. Analyze and experiment with forest regeneration of logged areas and the impact of different kinds of selective logging.

6. Study the reproductive biology and demography of important timber trees.

7. Study the feeding ecology of feral pigs in the forest, including experiments with fenced enclosures that keep pigs out.

8. Start gathering data on yearly stream flow, especially from both cleared and intact watersheds.

9. Start gathering more basic climatological data for both sides of the island, and higher elevations if possible.

10. Study daily cloud patterns in the mountains.

11. Continue taxonomic study of the flora and build ethnobotanical databases using local experts as resources.

12. Study the economic potential of the flora included in the reserve or buffer zone.

13. Survey the flora for chemical compounds of potential economic or medical importance.

Literature Cited

Conn, B.J. 1994. Documentation of the flora of New Guinea. *In*: C.I. Peng and C.H. Chou (eds.), Biodiversity and Terrestrial Ecosystems. Institute of Botany, Academia Sinica Monograph Series No. 14.

Johns, R.J. 1993. Biodiversity and conservation of the native flora of Papua New Guinea. *In*: B. M. Beehler (ed.), A Biodiversity Analysis for Papua New Guinea. Papua New Guinea Conservation Needs Assessment Vol. 2. Biodiversity Support Program, Washington, DC.

Paijmans, K. 1976. Vegetation of New Guinea. Melbourne University Press, Melbourne.

Peekel, P.G. 1984. Flora of a Bismarck Archipelago. Office of Forest Care, PNG.

Stevens, P.F. 1989. New Guinea. *In:* D.G. Campbell and H.D. Hammond (eds.), Floristic Inventory of Tropical Countries. New York Botanical Garden, Bronx, New York.

Chapter 2

A Transect-Based Floristic Reconnaissance of Southern New Ireland

Wayne Takeuchi and Joseph Wiakabu

Abstract

Floristic patterns are discussed from three 0.2 hectare transects established during the Rapid Assessment Program (RAP) biological survey of New Ireland's mountainous southern zone. The findings from transect-based inventory and *ad libitum* collecting indicate that humid forests in the surveyed tract are impoverished relative to mainland New Guinea. Species richness of canopy trees declines monotonically with elevation, but is also accompanied by increased percentages of Papuasian endemics within the montane communities. Nontree species exhibit similar patterns as overstory taxa, although the trends are less pronounced. Epiphytes are a particularly prominent component of the cloud-zone vegetation, accounting for more than one third of all taxonomic registers. Approximately 500 tracheophyte species were documented by the expedition, including a previously unknown orchid species in the endemic genus *Saccoglossum*, and five other novelties in *Corsia* (Corsiaceae), *Dendrobium* (Orchidaceae), *Freycinetia* (Pandanaceae), and *Psychotria* (Rubiaceae).

Introduction

Although Papua New Guinea (PNG) is widely recognized as a center for biotic diversification and endemism, many aspects of the Papuasian rainforest ecosystem are imperfectly understood. Basic themes in community structure, stand composition, species-area relationships, and elevational dependency of floristic variables still remain fruitful subjects for investigation. During the 1994 RAP survey, three strip-plots were completed on an altitudinal series from the Weitin River to the high ridges of the Hans Meyer Range. The transect activities were a taxonomically-focused adjunct to the general vegetation assessment presented by R. Foster in the preceding chapter (Chapter 1). A brief overview of the history of botanical exploration in New Ireland is also included in the Foster account.

Interpretation of botanical results from the 1994 RAP survey was complicated by the lack of information from earlier explorations. A 1975 Kew-sponsored expedition to the Hans Meyer Range yielded many significant discoveries, which were unfortunately not synthesized into a published account (but cf. Sands 1989). A subsequent PNG Forest Service visit also has no summary of results. However because of the primary evidence represented by the collective gatherings, southern New Ireland is currently one of the better-documented floristic districts in PNG.

The expedition's orchid specimens have been reviewed elsewhere (Howcroft 1994, 2001; Howcroft and Takeuchi in review) and new species in Corsiaceae, Pandanaceae, and Rubiaceae were previously described from the survey vouchers (Huynh 1999; Takeuchi and Pipoly 1998). Thus far, a total of six plant novelties have resulted from the expedition's collections.

General Description of Transects

The sample plots were sequentially established at expedition camps designated as the Weitin Base, Lake, and Top Camps. The transects had the following salient site attributes:

Transect 1, Weitin Base Camp. Near junction of the Niagara and Weitin Rivers, 152° 56' 242" E, 4° 30.210' S. Origin at 305 m asl (aneroid). Baseline 333 m long, ending at 373 m asl (aneroid). Extending on an azimuth of 320° magnetic. Topography: uniform incline, slope 12%. Situated on a low ridge flanked by feeder streams to the Niagara River. Above the flood/surge zone. Substrate of organic clay with an overburden of undecomposed litter. Surface rocks generally absent. Community type: lowland rainforest with scattered windthrow-induced gaps.

Transect 2, Lake Camp. Hans Meyer Range, NE of the Weitin River, 152° 56.489' E, 4° 27.205' S. Origin at 1175 m. Baseline 333 m long, ending at ca. 1190 m. Extending on an azimuth of 350° magnetic, approximately following the ridgecrest adjacent to Lake Camp. Average slope <10%. Clay substrate. Community type: pristine mossy forest, low montane by composition and physiognomy.

Transect 3, Top Camp. Hans Meyer Range, Angil Mountain, 152° 56.8' E, 4° 25.2' S. Origin at 1800 m. Baseline 333 m long, ending at ca. 1810 m. Extending on an azimuth of 0° magnetic. Topography: level ground on the ridgecrest next to Top Camp. Slippery clay substrate covered by bryophyte mats and ferny growth. Community type: pristine cloud forest with considerable epiphytic development.

Methods

Community parameters were assessed using conventional procedures in vegetation analysis. Protocols and conceptual guidelines as presented in Mueller-Dombois and Ellenberg (1974) are well-established standards, and served as models for the sampling.

At each study site, a reference baseline was extended by metric tape for 333 m from the point of origin. With the baseline as center axis, two contiguous files of 3 m x 10 m subplots were progressively advanced during data collection. Within the rectangular subplots, all trees with dbh ≥ 5 cm were counted and measured by diameter tape. Mapping was conducted using the baseline as a grid reference, in order to assess spatial patterns in tree recruitment and establishment. A total of 66 countplots was thus positioned, the final pair being extended an extra 3 m in order to enclose a sample tract of 2,000 sq m (0.2 ha).

All species within the transect were checklisted during the census. Records were vouchered by herbarium collections, with the exception of sterile individuals from distinctive taxa known with certainty to the writers. Attention was also directed to the growth status of the various species (i.e., fruticose/arborescent, nonwoody terrestrial, climber, and epiphyte) to estimate the relative representation of the principal growth strategies.

From transect enumerations, an estimate of floristic richness was directly obtained for each site. Conversion of the data to a simulated succession of nested plots allowed the number of taxa within territories of sequential extension to be determined. The species-area relationship was then graphically depicted for calculation of the minimal area representative of the community's composition (e.g., the Cain criterion).

At the Weitin Base Camp transect, a comparison was also made between line-sampling and the original contiguous-areas protocol imposed by strip plots. This action was undertaken to test the cost-effectiveness of different procedures for subsequent rapid assessments in Papuasia. Three parallel survey lines were placed through the same community as the first sample strip, either as superimposed or bracketing tracks (Figure 2.1). Censusing was then conducted at points spaced 5 m apart along parallel bearings, until an identical number of data replicates had been obtained as the plot counts. Since the same forest community was sampled, any contrast in the enumerations produced by each program could be attributed to the protocols themselves, and their respective effectiveness thus gauged.

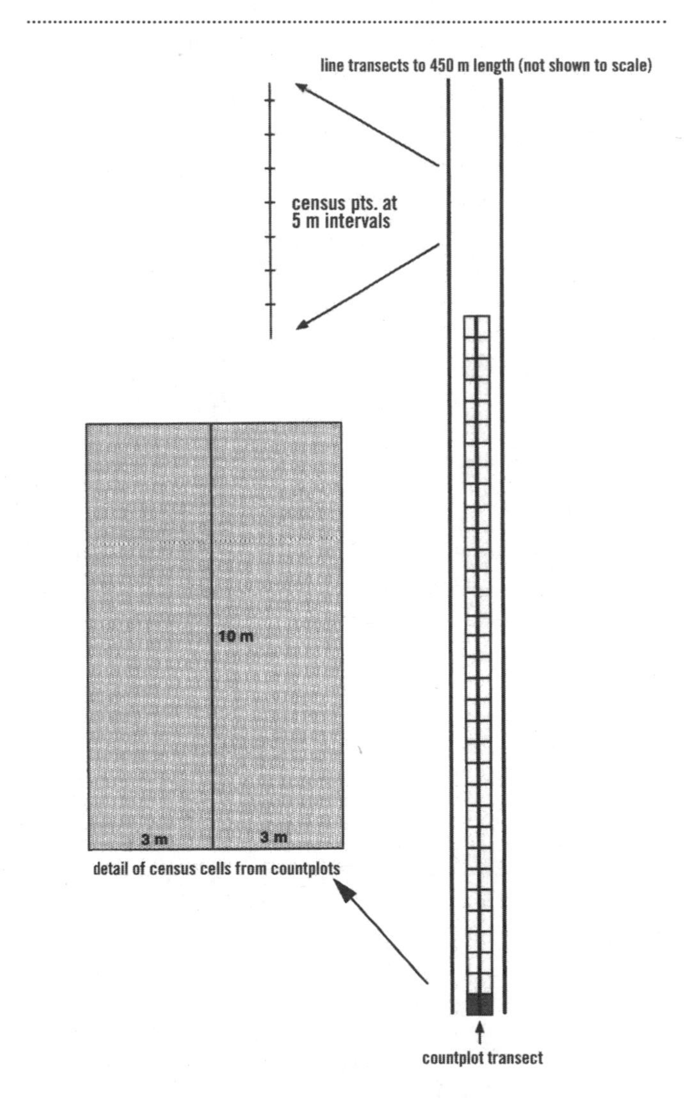

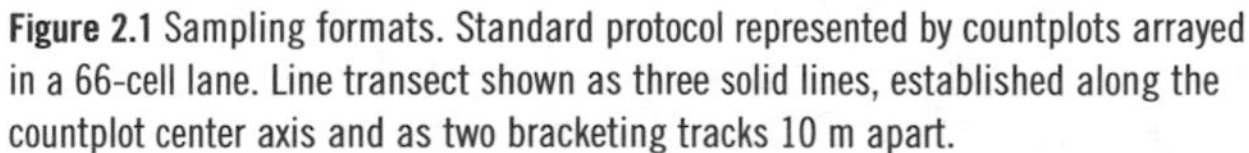
Figure 2.1 Sampling formats. Standard protocol represented by countplots arrayed in a 66-cell lane. Line transect shown as three solid lines, established along the countplot center axis and as two bracketing tracks 10 m apart.

As part of the general program of documentation, efforts were made to obtain herbarium specimens of every species encountered, regardless of whether the taxa occurred in transects or not. Collections were field-pressed and preserved with 70% ethanol for subsequent processing at Lae National Herbarium (LAE). The principal repository for survey vouchers is LAE. Residual duplicates are primarily at Kew (K), National Herbarium Nederland (L), and Harvard University Herbaria (A). The duplicate sets have frequently been allocated according to the presence of family specialists at the respective institutions.

Results

The transect localities represent perhumid or everwet forest (cf. McAlpine et al. 1983). Other floristic characteristics were determined as follows:

Transect 1, Weitin Base Camp

Closed canopy lowland rainforest with *Pometia pinnata* as the dominant canopy tree. *Neoscortechinia forbesii* was the frequency dominant among arborescent species. Understory open, with a herb layer consisting primarily of *Selaginella* cf. *durvillei*, *S. velutina*, rosette-stage *Calamus*, and various fern species. Lianes were physiognomically prominent, occurring as numerous ropes reaching into the canopy. Spiny-stemmed *Freycinetia* spp. were common in this latter group.

Transect 2, Lake Camp

Closed-canopy lower montane cloud forest, with *Syzygium* spp. the stature-dominant overstory component. Among trees and shrubs, *Spathiostemon javensis* was the frequency dominant. There was substantial development of the bryophyte and epiphyte flora. Understories were typically open.

Transect 3, Top Camp

Closed canopy montane mossy forest, with *Metrosideros salomonensis* as the dominant overstory species. *Ascarina* sp. (*maheshwarii* or *philippinensis*) and *Platea excelsa* were frequency co-dominants in the subcanopy. The understory was composed primarily of herbaceous taxa, especially orchids and ferns, which collectively comprised more than half of the nontree species.

The census data from all transects are summarized in Figure 2.2. Tree species satisfying the requirement of dbh ≥ 5 cm are tallied separately from the taxa not meeting this standard. The latter are designated as "NTS" (nontree species) and include all other plants encountered in the transect but not enumerated in the tree counts. A histogram of total richness is then obtained by summation of the preceding fractions. It is apparent that floristic diversity within the plots declines monotonically with elevation. In the preceding chapter (Chapter 1), Foster noted that richness of arborescent plants peaks at 750 m, but because our own plots were established at highly disjunct stations, we have no data from that elevation.

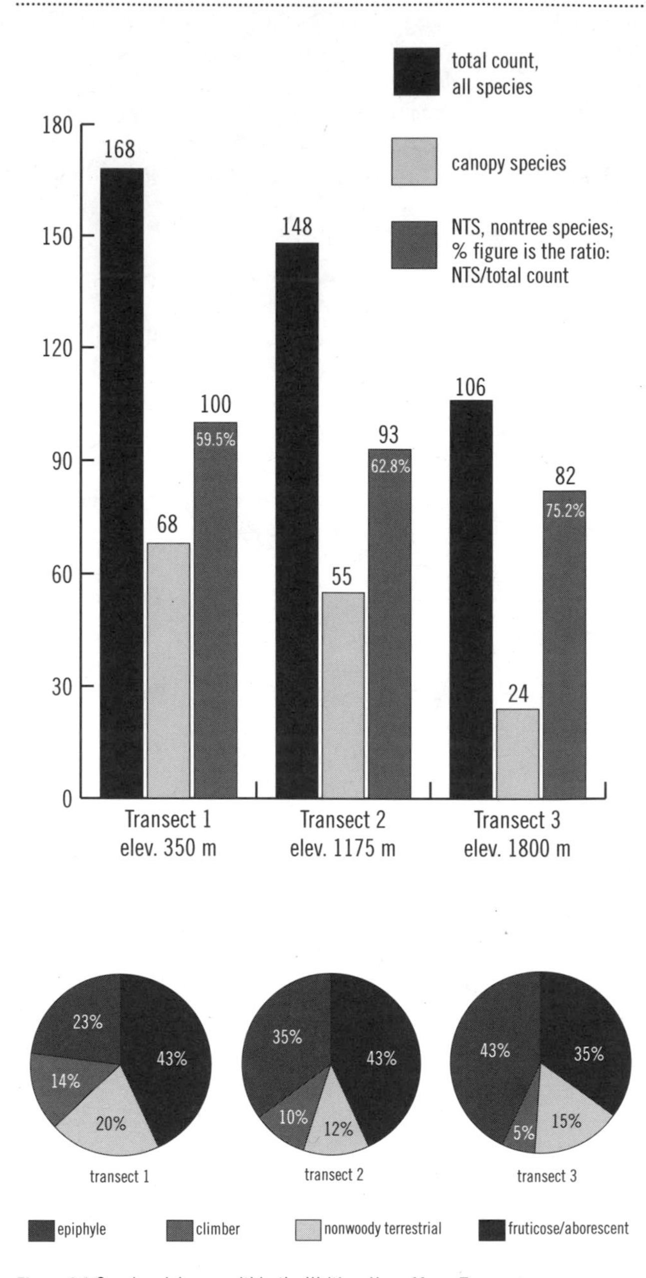

Figure 2.2 Species richness within the Weitin - Hans Meyer Transects.

In the present discussion, "richness" and "diversity" are regarded as equivalent concepts, using number of species as the relevant index (sensu Whittaker 1977). For purposes of rapid assessment, this is probably the most appropriate procedure, though subject to a dependence on sample size and time invested in search.

Otherwise, on Margalef's index, richness is 12.17 for Weitin Base and 8.81 for Lake Camp. Menhinick's index would be 4.34 and 2.57 respectively. Calculated values from the Shannon diversity function are 3.77 for Weitin Base and 3.37 for Lake Camp. Application of the historically earlier Simpson formulation yields corresponding values of .0296 and .0477. (The apparent reversal of numerical trend on the Simpson index is due to the way the index is defined, and the result is actually consistent with the other indices.) The Top Camp Transect cannot be evaluated on any of the preceding descriptors since individual frequencies could not be obtained within the limited time available at that site.

By any measure, at higher elevations there is clearly an overall floristic impoverishment, particularly in the canopy. Because of diversification in the epiphytic flora at altitude, suppression of nontree richness is minimal, and the apparent signs of a slight reduction within montane environments may not be real.

With increasing elevation there are also distinct changes in the growth strategy adopted by plants. Moving from the lower to higher elevations, the percentages of the different growth forms in the transects are: fruticose/arborescent taxa 43%, 43%, 35%; nonwoody terrestrials 20%, 12%, 16%; climbers 14%, 10%, 6%; and epiphytes 23%, 35%, 43%. The cloudy uplands are thus rich in epiphytes and poor in climbers, while the terrestrial species are somewhat constant across the investigated range. In general, the decrease in climbers with elevation is accompanied by a progressive increase in epiphytes.

A total of 690 collection numbers was obtained during the course of our survey. Approximately 400 taxa were identified to species level from sightings and exsiccatae. Many sterile gatherings were determinable only to genus. Due to an overall lack of pertinent data, distributional records were difficult to ascertain, but 101 taxa (54 from the low elevation sites and 47 from the montane camps) were preliminarily identified as possible range extensions (Takeuchi and Wiakabu 1994). These latter estimates, however, do not take into account the collections from the 1975 Kew expedition, which were not accessible in their entirety to the authors at LAE. From the list of determined species, a phytogeographic summary was also developed, indicating that percentage representation by Papuasian endemics increases from 31% to 74% in the transition from lowland to montane habitats (Takeuchi and Wiakabu 1994, part 3). Appendix 1 provides a tabulated summary of the botanical gatherings.

Discussion

The New Ireland transects provide a general measure of community composition and richness, and of the interrelation between these parameters with elevation. Species counts from our transects (Figure 2.2) are low by tropical forest standards (e.g., Gentry 1988, Paijmans 1970), and may reflect the existence of dispersal barriers between New Ireland and the Malesian source areas, as well as island-area effects sensu MacArthur and Wilson (1967).

In the montane transects from New Ireland, epiphytes account for 35–43% of all enumerated species (Figure 2.2), a proportion comparable to the highest values (35%) obtained for wet neotropical forests by Gentry and Dodson (1987a, 1987b). For purposes of documentation, the growth-form status assigned to the various taxa is shown as a coded listing in Appendix 1.

In Figure 2.3, the species/sampled-area relationship is charted for the three transects. Species-area accumulation declines with elevation. At Top Camp, the latent diversity has been adequately sampled at a minimum area of 400 sq m, and the curve is practically level. A sample tract of 400 sq m coverage should be sufficient to reveal an essentially complete suite of the canopy species for that community. In contrast, the 2,000 sq m area is clearly insufficient for sampling tree taxa from the Lake and Weitin Base communities. For these sites, the species-accumulation curves continue to climb with

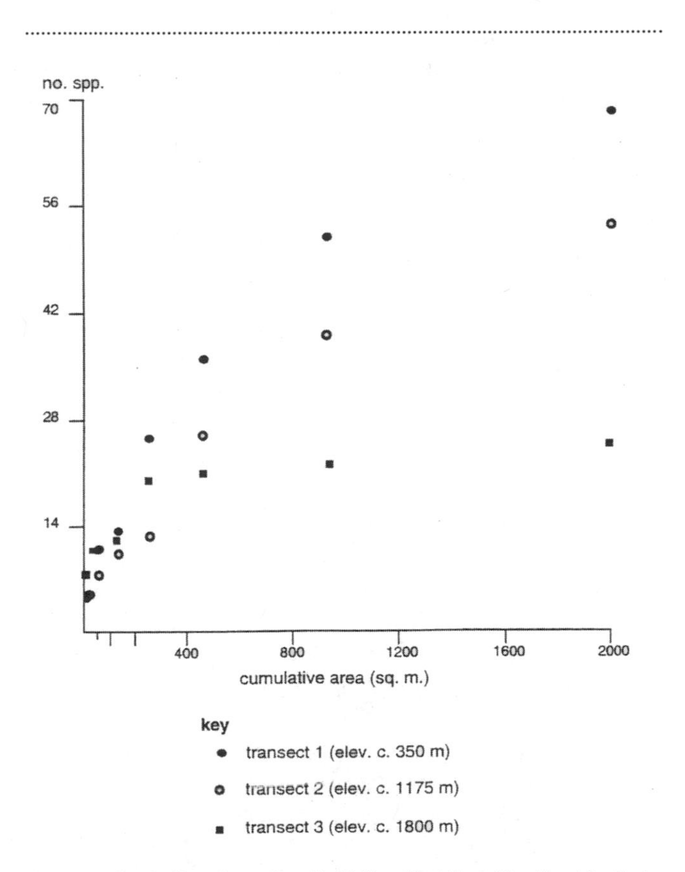

Figure 2.3. Species/Area Curves from the Weitin - Hans Meyer Elevational Gradient (Arborescent Taxa).

nearly unchanged positive slopes. Intensive surveys would be needed to determine the amount of additional effort required for saturation sampling of lowland and lower montane environments.

Prior to 1985 there were virtually no quantitative or plot-based studies of forest structure in Papuasia other than the widely cited account by Paijmans (1970). However in the last ten years a pronounced reversal of the previous neglect has occurred, with numerous investigations having been completed or currently underway (cf. Abe 2000; Abe et al. 1998, 1999, 2000; Balun et al. 1993, 1996; Conn 1985; Damas 1999, Damas et al. 1998; Kiapranis 1990, Kiapranis & Balun 1993; Kiyono et al. 1999; Kulang et al. 1997; Kuroh et al. 1999; Oatham & Beehler 1997, 1998; Oavika 1999; Reich 1998; Weiblen 1998; Wright et al. 1997). Contemporary investigators employ a variety of procedures and different threshold values for the tree enumerations, so comparisons are often problematic. However the studies by Balun et al. (1996) and Kulang et al. (1997) were based on the sampling protocols from the 1994 New Ireland RAP survey and are thus directly comparable to the results presented here. Procedures similar to the New Ireland RAP have also been adopted by ongoing work linked to the Pacific Asia Biodiversity Transect net (PABITRA: website at botany.hawaii.edu/pabitra), particularly by the Village Development Trust.

If the transect results from the present study are combined with the Madang inventory by Kulang et al. (1997), and with the New Britain survey by Balun et al. (1996), the resulting patterns are generally in conformity to expectations from MacArthur-Wilson theory. The mainland PNG environment gives the highest species counts in altitudinally sequenced samples (Kulang et al. 1997), and the Bismarck archipelagic stations of New Britain and New Ireland are distinctly depauperate at corresponding elevations. The impoverishment in the two Bismarck localities is especially acute in the montane zone, where it has been hypothesized that diversification is at an incipient stage due to geological recency of the cloudy upland habitat (Takeuchi & Pipoly 1998). One disconforming element is the substantially lowered counts from New Britain relative to the smaller territory of New Ireland, suggesting the importance of local factors. Elevation-based assessments in the future should focus specifically on the gap from ca. 600–1000 m. There is accumulating evidence that the point of highest botanical richness lies somewhere in that interval (cf. Takeuchi & Golman 2001).

In Papuasian forest communities, many canopy species often exhibit contagious distributions, with distinctly aggregate rather than even occurrences (pers. obs.). Such clustering is suggestive of the presence of cohorts, in which individuals establish at a common point in time in response to some forest dynamic phase like gap formation. Natural seeding of the ground around a mature tree is also likely to produce conspecific aggregates. Plot-based transects, by their very nature, will tend to intersect and encompass trees establishing in such patch-wise fashion. Enumerations would be skewed by repetitive counting, thus lowering the efficiency of the sampling process. An attempt was made to determine the extent to which these expected effects occur, using separate line and plot transects in side by side fashion as described earlier.

Results from the comparative sampling show an unexpected equivalence. Enumeration efficiencies (countplots: 68 canopy species recorded from 246 individuals; versus line sampling: 69 canopy species recorded from 246 individuals) are virtually identical. The similarity in outcome was initially attributed to the overall impoverishment of the flora. In a species-poor situation, differing methodologies can be expected to produce equivalent results since the opportunities for variance would be dampened by the limited number of species. However the taxa recorded in common by either format (i.e., 46 spp.) was barely greater than the total of 45 species uniquely registered by one or the other program. Such an apportionment is hard to reconcile with the notion that diversity had been limiting; otherwise the number of commonly recorded species should have far exceeded the second figure.

In retrospect the two methodologies should have been applied on more divergent protocols. The plot could be established as a square block, advanced on a broad front (rather than only 6 m) in order to maximize the effects caused by sampling over continuous areas. Line enumerations could be similarly more open-interval (e.g., censusing at points separated by ≥10 m), to optimize effects resulting from spatial separation. But within the parameters established for the dual sampling actually conducted, there are no distinctions to be made in outcome. If a sampling efficiency curve (Figure 2.4) is constructed for the respective formats, they virtually superimpose on each other. Even the distribution of counts among the different taxa recorded (Figure 2.5) produces no essential departures. The line sampling gives more single-occurrence records than the countplot, but the patterns are equivalent. The statistical spread of common and infrequent species is the same on each procedure.

Even if initial expectations were not supported, this aspect of the investigation has practical consequences for future rapid-assessment surveys. The plot transect required more than twice as much time and labor as the line sampling. If closely contrasting methods do not produce diverging results, the easier one should be favored. When the objective is a simple diversity or richness assessment, there is little to be gained with cost-intensive plots.

Acknowledgments

The text incorporates many suggestions and comments from B. Beehler. Prof. emeritus D. Mueller-Dombois also corrected the draft.

A number of taxonomic specialists made plant identifications or assisted indirectly by generously sharing information on their respective groups. We are indebted in this fashion to B.A. Barlow (Loranthaceae), B.L. Burtt (Gesneriaceae), O. Gideon (Costaceae), H. Fortune-Hopkins (Cunoniaceae), N.H.S. Howcroft (Orchidaceae), J. Pipoly III (Myrsinaceae), P. F. Stevens (Clusiaceae, Ericaceae, passim), and G. Weiblen (Moraceae).

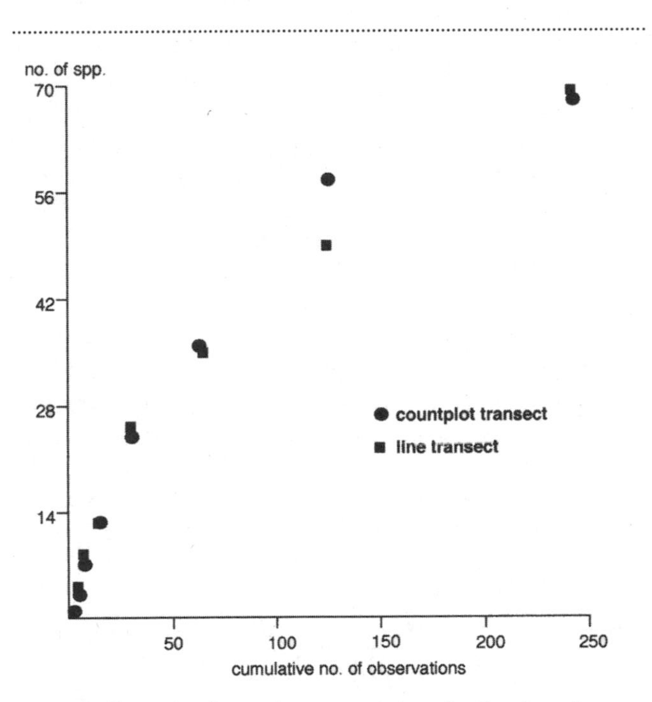

Figure 2.4. The number of canopy taxa enumerated as a function of sampling intensity.

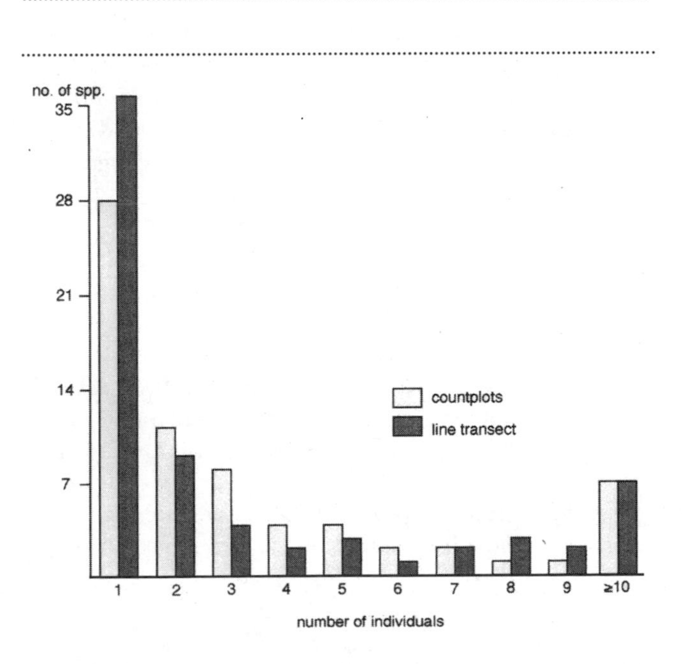

Figure 2.5. Profile of the number of individuals per species, as determined by each of the sampling formats.

Literature Cited

Abe, H. 2000. Effect of logging on forest structure at the Mongi-Busiga forest research plots, Finschhafen, Papua New Guinea. PNGFRI Bull. 18: 66–77.

Abe, H., N. Sam, M. Niangu, K. Damas, and Y. Kiyono. 1998. Introduction to research of effect of logging on lowland tropical rain forest at Finschhafen in Papua New Guinea. JICA-PNGFRI Forest Research Note 10.

Abe, H., K. Damas, M. Niangu, N. Sam, and Y. Kiyono. 1999. An enumeration and species diversity study at Mongi-Busiga forest research plots in Finschhafen. PNGFRI Bull. 12: 96–105.

Abe, H., N. Sam, M. Niangu, K. Damas, P. Vatnabar, Y. Matsuura, Y. Kiyono. 2000. Preliminary results of the study on the effect of logging at Mongi-Busiga, Finschhafen, Papua New Guinea. PNGFRI Bull. 17: 1–79.

Balun, L., A. Vinas, and L. Orsak. 1993. Study on the biological diversity of Hunstein Range, East Sepik Province, Papua New Guinea. *In*: R. Höft (ed.), Proceedings of the Biological Society of New Guinea. Annual Meeting 1993. Wau Ecology Institute, 9–26.

Balun, L., Emrik, and L. Orsak. 1996 (unpublished). A study on plant species diversity and spatial patterns in rain forest communities from Sulka area in New Britain Island, Papua New Guinea. Pacific Heritage Foundation, 38 pp.

Conn, B.J. 1985. The Watut/Wafi expedition, Papua New Guinea. 9th February to 7th March 1985. National Herbarium of Victoria, 14 pp. with 5 appendices.

Damas, K., H. Abe, N. Sam, M. Niangu, and Y. Kiyono. 1998. Tree species composition of Busiga plots. JICA-PNGFRI Forest Research Note 10.

Damas, K. 1999. Residual tree species composition in logged-over forests. PNGFRI Bull. 12: 35–46.

Gentry, A. 1988. Changes in plant community diversity and floristic composition on environmental and geographical gradients. Ann. Missouri Bot. Gard. 75: 1–34.

Gentry, A., and C. Dodson. 1987a. Contribution of non-trees to species richness of tropical rain forest. Biotropica 19: 149–156.

Gentry, A., and C. Dodson. 1987b. Diversity and phytogeography of neotropical vascular epiphytes. Ann. Missouri Bot. Gard. 74: 205–233.

Howcroft, N.H.S. 1994. Orchid collections from New Ireland. Orchid Research Bulletin 3. Papua New Guinea Forest Research Institute.

Howcroft, N.H.S. 2001. A new species of *Saccoglossum* (Orchidaceae) from the Hans Meyer Range, New Ireland, Papua New Guinea. Sida 19 (3): 519–521.

Howcroft, N.H.S., and W. Takeuchi (in review). New and noteworthy orchids from the Bismarck Archipelago, Papua New Guinea. Sida.

Huynh, K.-L. 1999. The genus *Freycinetia* (Pandanaceae) in New Guinea (part 2). Bot. Jahrb. Syst. 121: 149–186.

Kiapranis, R. 1990. Plant species enumeration in a lowland rain forest in Papua New Guinea. *In:* D.A. Taylor and K.G. MacDiken (eds.), Proceedings of International Workshop held on 19–29th November 1990. Winrock Int. Inst. Agric., Los Banos.

Kiapranis, R., and L. Balun. 1993. Forest structure of a montane rain forest in Wau, Morobe Province, Papua New Guinea. *In*: R. Höft (ed.), Proceedings of the Biological Society of New Guinea. Annual Meeting 1993. Wau Ecology Institute, 3–7.

Kiyono, Y., M. Niangu, N. Sam, K. Damas, and H. Abe. 1999. Preliminary studies on structure of a lowland evergreen forest at Finschaffen, Morobe Province, Papua New Guinea. PNGFRI Bull. 11: 46–68.

Kulang, J., O. Gebia, and L. Balun. 1997 (unpublished). A study on the biological diversity of the Hagahai area in the Madang Province, Papua New Guinea, 47 pp.

Kuroh, B., S. Musa, and C.P. Phee. 1999. Status of logged-over forest of Papua New Guinea. PNGFRI Bull. 12: 13–25.

McAlpine, J.R., and G. Keig, with R. Falls. 1983. Climate of Papua New Guinea. CSIRO and Australian National University Press, 200 pp.

MacArthur, R.H., and E.O. Wilson. 1967. The theory of island biogeography. Princeton University Press, 203 pp.

Mueller-Dombois, D., and H. Ellenberg. 1974. Aims and methods of vegetation ecology. John Wiley and Sons, New York.

Oatham, M., and B.M. Beehler. 1997. Richness, taxonomic composition, and species patchiness in three lowland forest treeplots in Papua New Guinea. Proceedings of the International Symposium for Measuring and Monitoring Forests and Biological Diversity; the International Networks of Biodiversity Plots. Smithsonian Institution/Man and the Biosphere Biodiversity Program (SI/MAB).

Oatham, M., and B.M. Beehler. 1998. Richness, taxonomic composition, and species patchiness in three lowland forest treeplots in Papua New Guinea. *In*: F. Dallmeier and J.A. Comiskey (eds.), Forest Biodiversity Research, Monitoring and Modeling. Conceptual Background and Old World Case Studies. Man and the Biosphere Series, vol. 20. UNESCO and the Parthenon Publishing Group, Paris and New York, 613–631.

Oavika, F. 1999. Growth of forests after logging in Papua New Guinea. PNGFRI Bull. 12: 26–34.

Paijmans, K. 1970. An analysis of four tropical rain forest sites in New Guinea. Journ. Ecol. 58: 77–101.

Reich, A. 1998. Vegetation part 1: A comparison of two one-hectare tree plots in the Lakekamu basin. *In*: A. Mack (ed.), A biological assessment of the Lakekamu Basin, Papua New Guinea. Rapid Assessment Program Working Papers no. 9, Conservation International, Washington, DC, 25–35, 97–104.

Sands, M.J.S. 1989. Botanical exploration in Papua New Guinea. *In*: F.N. Hepper (ed.), Plant Hunting for Kew. Her Majesty's Stationery Office, 103–116.

Takeuchi, W., and J. Wiakabu. 1994 (unpublished). A botanical report of the 1994 New Ireland expedition. Report submitted to the PNG Forest Research Institute, 251 pp.

Takeuchi, W., and J. Pipoly III. 1998. New flowering plants from southern New Ireland, Papua New Guinea. Sida 18 (1): 161–168.

Takeuchi, W., and M. Golman. 2001. Botanical documentation imperatives: some conclusions from contemporary surveys in Papuasia. Sida 19 (3): 445–468.

Weiblen, G. 1998. Composition and structure of a one hectare forest plot in the Crater Mountain Wildlife Management area, Papua New Guinea. Science in New Guinea 24: 23–32.

Whittaker, R. H. 1977. Evolution of species diversity in land communities. Evol. Biol. 10: 1–87.

Wright, D., J.H. Jessen, P. Burke, and H.G. de Silva Garza. 1997. Tree and liana enumeration and diversity on a one-hectare plot in Papua New Guinea. Biotropica 29 (3): 250–260.

Chapter 3

A Preliminary Assessment of the Insects of Southern New Ireland, with Special Focus on Moths and Butterflies

Larry J. Orsak, Nick Easen, and Tommy Kosi

Abstract

During entomological field surveys conducted in southern New Ireland during January–February 1994, we made sample collections along an elevational transect from the lowlands to 1830 metres above sea level. As a primary focus to these studies, nocturnal light-trapping was used to collect more than 16,000 moths of 1088 species.

The surveys delineated distinct moth assemblages at 250 m, 1180 m, and 1830 m. Overall, the size of the moth fauna was shown to be considerably poorer than that from a comparable transect in mainland New Guinea. By contrast, species richness at each sample site was higher than expected for an island of New Ireland's size, perhaps reflecting New Ireland's biogeographical linkage to the Bismarck and Solomon archipelagoes. Peak species richness apparently occurred between 300 and 1100 metres, a zone that was inadequately surveyed because of topography. Disturbed forest was shown to support a moth fauna poorer than that of a nearby undisturbed forest.

A new species of swallowtail butterfly, *Graphium kosii*, was discovered in the highland mossy forest. This, along with the birdwing butterfly *Ornithoptera priamus urvilliana* and several large beetles, best exemplifies the insects from southern New Ireland that show commercial potential.

Introduction

An ecological and commercial assessment was made of the insect fauna of the Weitin watershed of southern New Ireland. The commercial assessment sought to determine what insect species might be collected or "farmed" and sold through local marketing venues to collectors. The collection and rearing of butterflies and other insects could become a minor but stable forest-based source of revenue to the customary landowners.

The ecological assessment generally focused on Lepidoptera (butterflies, moths and skippers), but most effort was devoted to the Heterocera (moths). Moths were chosen because they show promise as overall biodiversity indicators. Predominantly plant-feeding in the larval stage, they may serve as a ready taxonomic probe into local and regional biodiversity. It has been shown elsewhere that insects show a high correlation with the spatial, architectural, and taxonomic diversity of plant communities (Southwood et al. 1979).

A tremendous advantage of using moths is the ease and speed with which a representative sample can be taken. In the tropics, a single sheet illuminated by a mercury vapor light can attract hundreds of species in a night; the single-night-trapping-record for PNG is 700+ species (Orsak *unpubl. data* from Tari Gap, Southern Highlands Province). This level of richness is useful in rapid biodiversity assessments, because with such a large fauna the presence or absence of a few species will not disproportionately influence survey results (as can happen with the less species-rich groups).

Holloway (1976, 1984) pointed out the attributes of moths as bioindicators from his work in Borneo, identifying groups (subfamilies, families) that seemed especially sensitive indicators of habitat change. The main problem in developing his findings into a widely-used technique is that his thrust was taxonomic—the moths first would have to be identified. Such identification work requires specialized training and often specialized equipment such as a dissecting microscope for dissection of genitalia.

An alternative is to forego identifications and simply sort the night's catch into what look to be species units—"morphological species." This was the strategy used in this study. We generally followed the methodology of Hebert (1980) and Thomas and Thomas (1994). From these "morpho-species" data, biodiversity indices were calculated, and the results used to extrapolate overall species richness at a site.

Methods

Collections were made in January–February 1994 from four general sites between 200–1830 meters elevation, corresponding to the expedition's camp sites. They were as follows:

1A. Weitin Base Camp, 250 m: Alluvial plain forest in the Weiten Valley, with obvious, considerable natural disturbance due to periodic flooding of the area. *Casuarina* was common nearby. Our moth collecting site was in a small dry stream clearing.

1B. Weitin Base Camp, wash 300 m: Lowland rainforest, 10% hillslope, dominated by *Pometia pinnata*. Bisected by a running stream. Our moth collecting site was on a forested hillslope overlooking the stream.

2. Lake Camp, 1200 m: Naturally forested ridge, surrounding a small natural lake. Tree flora dominated by a *Syzygium* species. Epiphytes and bryophytes abundant. Moth collecting sites were at lakeside, and 10 meters from the lake, facing into the forest.

3. Summit ("Top Camp"), 1800 m: High elevation, undisturbed mossy forest, with heavy epiphyte load. Canopy dominated by *Metrosideros salomonensis*, with subcanopy dominants *Ascarina* sp. and *Platea excelsa*.

4A. Stream ("Riverside Camp"), 200 m: Low elevation, previously heavily logged flat forest. Moth collecting site was in the dry stream bed.

4B. Stream Overlook, ("Riverside Camp") 205 m: Remnant unlogged site within the overall logged area, with the moth collecting site overlooking a stream.

Sampling for Commercially Valuable Insects

Butterflies and beetles are the most commercially valuable Papua New Guinean insects overall, with some walkingsticks also having some monetary value. We made casual, sporadic collecting efforts for these, largely using the services of our porters and village assistants on the expedition, mainly at the two lowest elevation sites. Butterflies were collected by net; beetles and walkingsticks were largely collected by hand, but sometimes also by net (i.e., day-flying beetles).

The specimens were mounted, then brought to the Insect Farming and Trading Agency (P.O. Box 92, Bulolo, Morobe Province) for an economic assessment. The Agency sells about K500,000 worth of specimens every year to overseas collectors, naturalists, museums, researchers, and artists.

Moth Diversity Assessments

Our usual light trap consisted of several white sheets illuminated by four 15-watt blacklight fluorescent bulbs and one unshielded 175-watt mercury vapor light. The sheets reflect light into the forest and serve as an attractive perching site from which the fieldworkers remove the moths for enumeration. The setups were established in a natural or artificial clearing, and shielded from rain by a slightly translucent blue plastic tarpaulin. More specifically, the clearings comprised a lakeside (Site 2), a stream (sites 1B, 4A), a wash (Site 1A), a clearing and artificial clearings in the forest (sites 2, 3, 4B). At the summit site (1830 m), the failure of the generator forced the use of a gas pressure lantern as the light source.

Lights were turned on at 6:30 pm to capture possible crepuscular fliers, such as hepialids (ghost moths). Collecting was continuous until 2:30 am. The latter time was chosen to allow the traps to attract what are exclusively late night-flying moths, such as the Hercules moth (*Coscinocera hercules*), which usually appears at the lights around 0100–0200 hrs.

An effort was made to collect every moth attracted to the sheet. Moths were placed in ethyl acetate killing jars, and after death, were stored in glassine envelopes. The next day, they were sorted into initial species-groups and repackaged. At the Base Camp, a preliminary "voucher collection" was initiated, consisting of one specimen of each morpho-species, labeled with a unique number.

As the sorting of moths continued at the Christensen Research Institute in Madang (PNG), the partially sorted moths were re-sorted into final morpho-species groups by several Papua New Guinean parataxonomists as well as Peace Corps volunteers. Sorting was accomplished by dumping out the contents of each envelope and further sorting the contents, until all the morpho-species had been segregated, then each of the morpho-species was matched up with a like specimen in the voucher collection. If a morpho-species was new to the voucher collection, a sample was pinned and added, and a new number was assigned to it. Counts of each morphospecies were computed for each site and entered into a computer spreadsheet.

The moth sorters were instructed not to lump slightly different looking specimens into the same morpho-species, but to split them as finely as necessary to reflect any consistent wing pattern or size characteristics. This was vetted against a large photographic collection of identified Papua New Guinea moths (including distinct infraspecific morphs of variable species) from the Australian National Insect Collection (Canberra) and the Bernice P. Bishop Museum (Honolulu, Hawaii, USA) for the diagnosis used for this analysis. With these photographs, the numbers and the specific moths in the voucher collection were amalgamated, so that only one representative number and specimen ended up for each species, no matter how variable that species was. The resultant collection reflected as much as possible the species concepts of current specialists.

With the information on species and frequency of each species for each night's collection, biodiversity indices could then be calculated.

Diversity Measures

The log series model describes a pattern of species abundance (Magurran 1988), and fits well with what has been found in many insect populations (Southwood et al. 1979). It gives heavy weighting to those species with medium abundance in the community, rather than the very abundant and very rare species. The log series distribution has two defining population parameters for a multispecies community, the most useful for the type of moth survey described here is *alpha*.

Alpha (α) is the parameter that is independent of sample size (when the number of individuals collected is > 1000) and characterizes the required population quality (Kempton and Taylor 1974). The application of the *alpha index* as a measure of "species richness" was first described by Fisher (1943). This index of diversity behaves consistently within a stable population and responds to changes within and differences between environments. Invariably replicate collections taken from a single habitat type give similar *alpha* indices (Hebert 1980). The log series takes the form of:

$$\alpha x, \alpha x^2/2, \alpha x^3/3, \ldots \alpha x^n/n$$

αx being the number of species predicted to have one individual, $\alpha x^2/2$ those with two and so on. The number of species, S, is obtained by adding all the terms in the series. However, as S and N (the total number of individuals collected) are already known, all that needs to be calculated is the value of x from the iterative solution (1) and then alpha, which is a function of both x and N (2).

$$S/N = (1 - x/x)(-\ln(1 - x)) \quad (1)$$

$$\alpha = N(1 - x)/x \quad (2)$$

Taylor *et al.* (1976), using 10 years of light-trap data at one site, found that the site's environmental stability was better reflected by *alpha* than alternative nonparametric indices (Simpson-Yule and Shannon Index). Thus, the *alpha* statistic is a very useful indicator for species richness, and was adopted for this study.

Results

Moth Biodiversity Assessments

16,194 moths of 1088 species were sorted from the New Ireland collecting sites. Table 3.1 summarizes the sampling effort, the number of specimens collected per site, the number of species found at each site, and the greatest number of species found on any one night. The largest collections came from the most intensively collected sites. Number of specimens collected per hour fell with increasing elevation.

Most species (Figure 3.1) were members of the families Pyralidae *sensu lato* (426 species), Geometridae (243 species), Arctiidae (104 species), and Noctuidae (116 species). Figure 3.2 shows the breakdown by site for these four dominant families. There are no changes in percentage composition with elevation, except that minor families encompass a greater

Table 3.1. Moth Capture Statistics for Four Collecting Sites in Southern New Ireland.

Site Number and Characteristics	No. Hours Sampled	No. of Specimens Collected	No. of Specimens per Hour	No. of Species Collected	Species Count on Best Night
Site 4A					
Riverside Camp; 200 m *Logged*	16	2815	176	313	185
Site 4B					
Riverside Camp; 205 m *Undisturbed* stream overlook	16	1441	90	326	202
Site 1A					
Weitin Base Camp; 250 m *Naturally disturbed* forest	24	1815	76	415	231
Site 1B					
Weitin Base Camp wash; 300 m *Undisturbed*	56	4903	88	721	350
Site 2					
Lake Camp; 1180 m *Undisturbed*	87	4274	49	690	450
Site 3					
High (Top) Camp; 1830 m	31.5	946	++	219	107

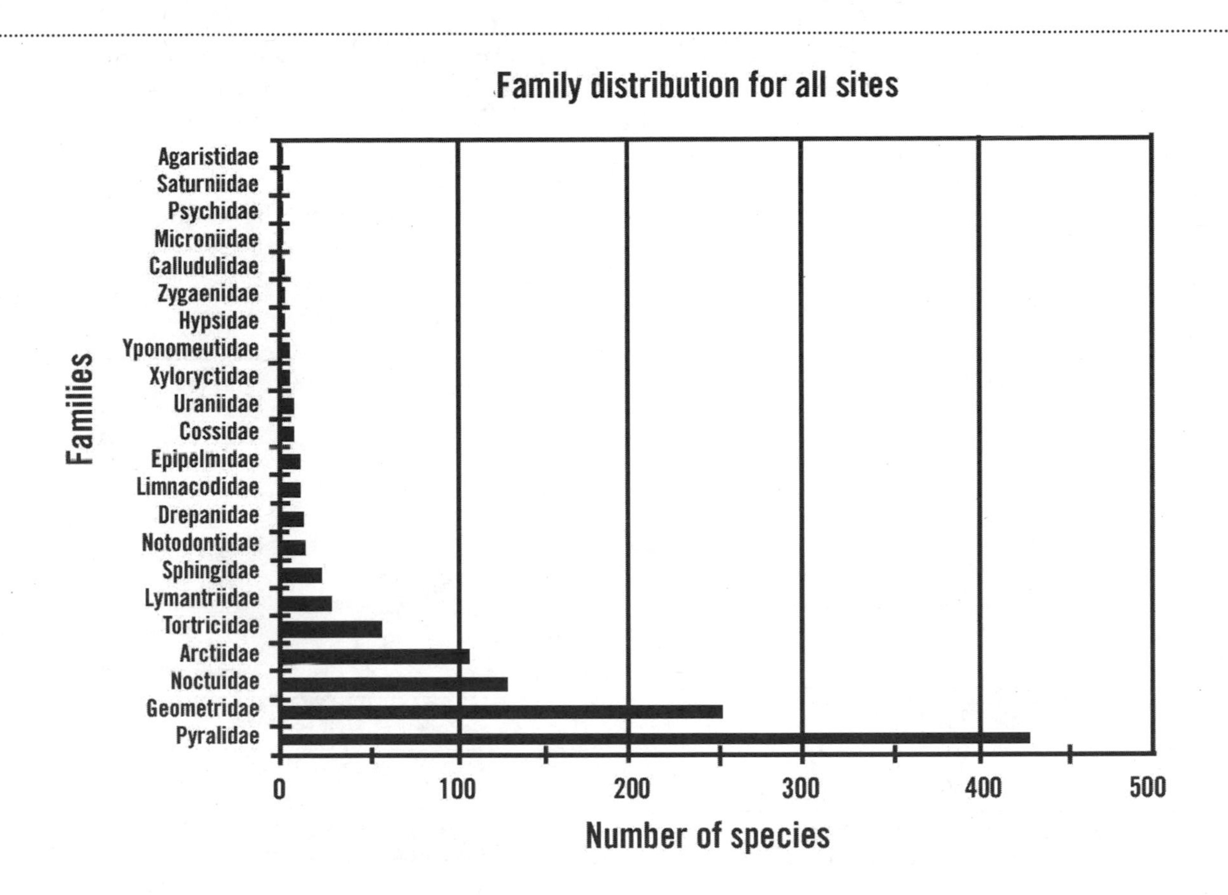

Figure 3.1. Relative dominance of each moth family in the overall southern New Ireland moth fauna as surveyed. The findings generally replicate those found elsewhere in the Southeast Asia mainland and the islands.

percentage of the overall moth community at lower elevations, and the Geometridae and Noctuidae seem to dominate more at higher elevations. Noctuids show a similar change with elevation in Borneo and Sulawesi (Holloway 1987).

Figure 3.3 shows the Rank-abundance plots for all the collection sites. All sites roughly fit the log series model, with a large number of species of low abundance and a low proportion of high abundance, with proportionate grades in between. But Site 2 (Lake Camp) data shows the lowest *Chi-Square* value and hence the closest fit. The *Chi-Square* values for the different sites were: 1A=40.3; 1B=16.9; 2=8.0; 3=14.8; 4A=28.1; 4B=40.6. *Chi-Square* values were lowest for the sites where the most specimens were collected, i.e., Sites 1B and 2. However, in general tropical populations deviate from a log series curve because of the preponderance of species for which only single specimens are found.

This is not the first time an excess of rare species has been noted (Hebert 1980, Kempton 1975) causing the observed results to deviate from the log series, thus producing overtly high *Chi-Square* values. But *alpha* itself not always strongly affected by such deviations (Taylor, Kempton and Woiwod 1976), though it is hard to quantify when the number of rare species with low abundance could begin to reduce the effec-

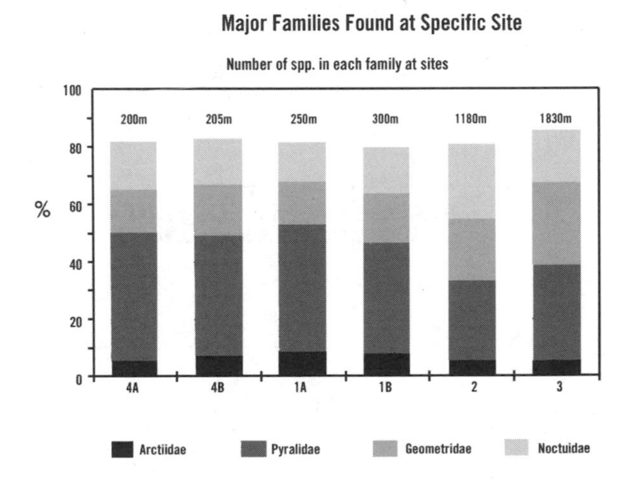

Figure 3.2. This shows the abundance of moth species within the main four families for each of the sites surveyed. It can be seen that there are some distinct trends, namely, the low number of Pyralids and the high number of Noctuids at the most diverse site (Site 2), as well as the gradual increase in the number of Geometrids as altitude increases. It is also interesting to see the small peak in Arctiids at 250m (Site 1B).

tiveness of *alpha* as a diversity index. Holloway's (1987) data shows that the fit is close enough and that *alpha* still remains a useful index to describe species richness.

Figure 3.4 graphs and compares *alpha* from each of the sites. From these values, the following trends emerge:

(1) *Alpha* increases with elevation up to somewhere between the base camp wash (Site 1B) and the Lake Camp (Site 2), and then decreases with elevation.

(2) *Alpha indices* were lower in disturbed areas, compared to nearby undisturbed areas. This was irrespective of

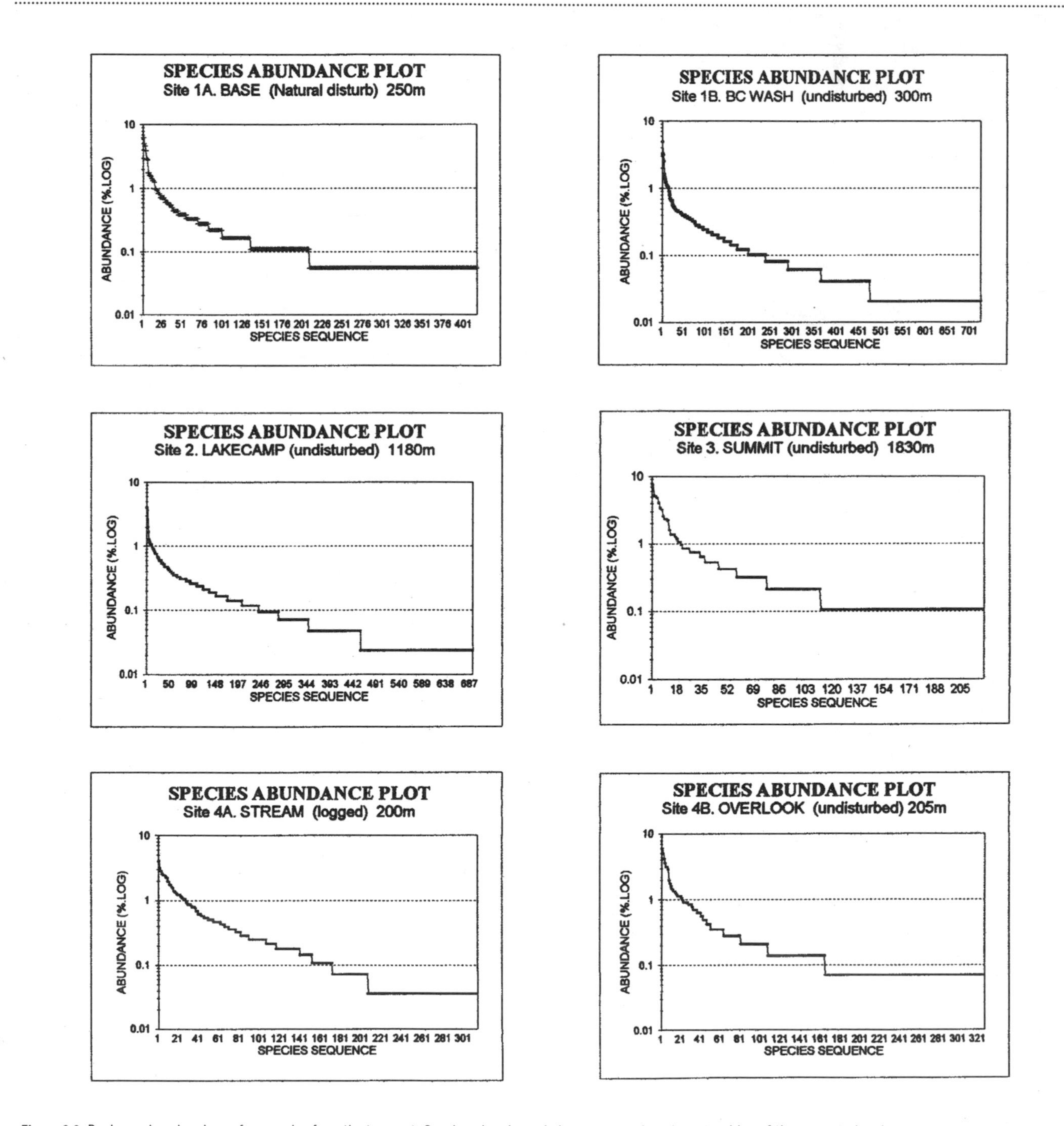

Figure 3.3. Rank species abundance for samples from the transect. Species abundance is here expressed as the natural log of the percent abundance.

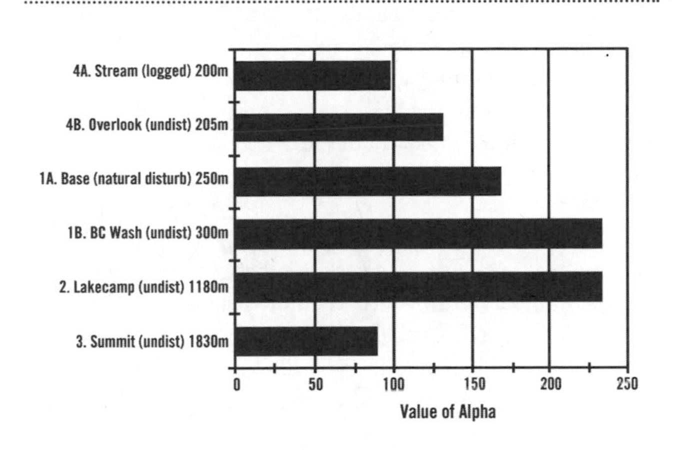

Figure 3.4. Summary of *Alpha* values for the samples from the six main sample sites.

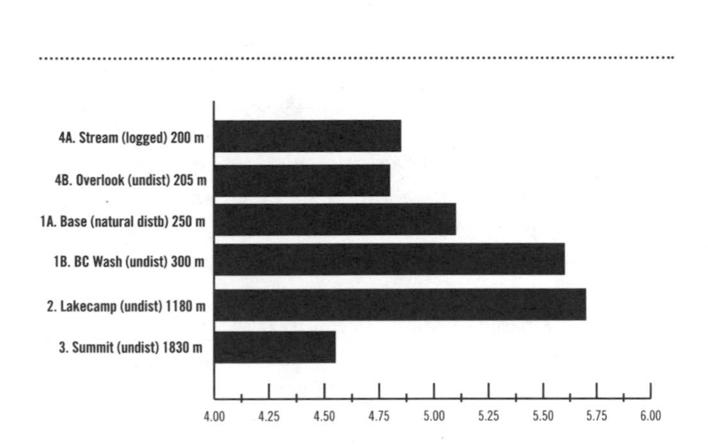

Figure 3.5. Shannon Index measures for the samples from the six main sites.

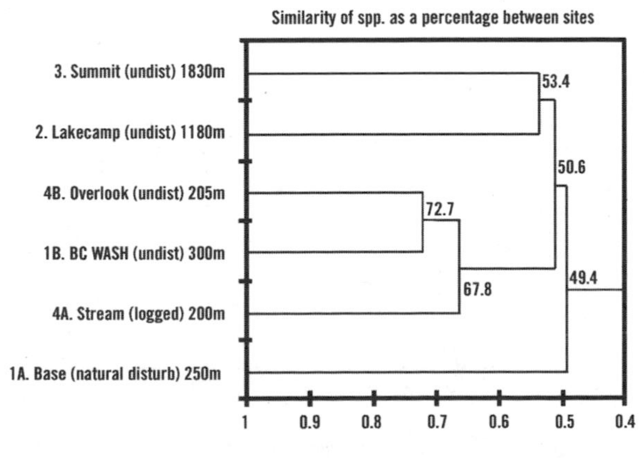

Figure 3.6. Dendrogram of similarity of species lists from the six survey sites.

whether the disturbance was natural (the alluvial forest near the base camp, Site 1A) or human-induced (severe logging at Site 4A).

The Shannon Index is a common, nonparametric statistic used to gauge and compare species richness (Magurran 1988). When this statistic is applied to our dataset, the resultant patterns closely follows that for *alpha* (Figure 3.5).

Evenness values range from 0.836 to 0.878, where 1.0 represents a situation where all species are equally abundant. It can be seen that the values are similar and high. This again reiterates the presence of many rare species with equal abundance.

Species Similarity Between Sites

The dendrogram in Figure 3.6 illustrates similarities in the moth communities at different sites. The highest degree of overlap (73%) occurs between Sites 4B (elev. 205 m; undisturbed forest) and 1B (elev. 300 m; undisturbed forest), which is reasonable considering how close the two sites are in elevation. The two "cloud forest" camps, at 1180 m and 1830 m, share over 50% of their species. The two disturbed sites (4A, 200 m, logged; and 1A, 250 m, natural disturbance) each had a distinct community compared to their nearest undisturbed collecting site counterparts.

Although Sites 1 (Weitin Base Camp) and 4 (Riverside Camp) were of similar elevation, the communities were nonetheless distinct, both for the disturbed forest collecting sites and the undisturbed sites. In both cases, a number of moth species found down towards the valley floor, closer to the coast, did not extend into the valley; in contrast, "up valley" moth species were more likely to extend to down-valley sites.

A comparison of "single-site" species is graphed in Figure 3.7—the percentage of *all* species collected during the survey that were found at only one of the collecting sites. This likely

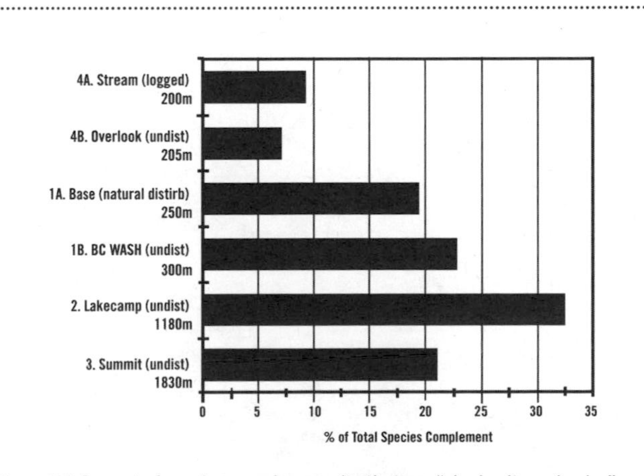

Figure 3.7. Percent of species samples per site that are "single site endemics"—recorded during the survey from just a single site.

is a correlate of sampling effort (compare trend in Figure 3.7 with sampling effort data in Table 3.1).

Green Geometrines as a Biodiversity Indicator

Undisturbed, high elevation mountain sites in Papua New Guinea contain an array of delicate, often translucent-winged green moths in the subfamily Geometrinae (*Anisozyga* spp., *Thalassodes spp., Prasinochyma spp., Gelasma* spp. etc.). In Wau, disturbed forests contain considerably fewer of these species than undisturbed forests in the surrounding mountains (Orsak unpubl. data). Thus, it has been suspected that green geometrine moths might be good indicators of overall moth species richness.

Figure 3.8 summarizes the *alpha indices* for green geometrine moth species. It was felt that separate *alpha* indices should be calculated for the green geometrine moth species, because variable numbers of species and individuals were collected at the different sites, thus some sort of diversity measure was useful. It can be concluded that species richness was highest at the Lake Camp (1180 m), though the green geometrine moths made up a higher *percentage* of the total moth community at the Top Camp (1830 m). The two highest sites contain the highest percentages of green geometrines; these were also the sites with the highest epiphyte load.

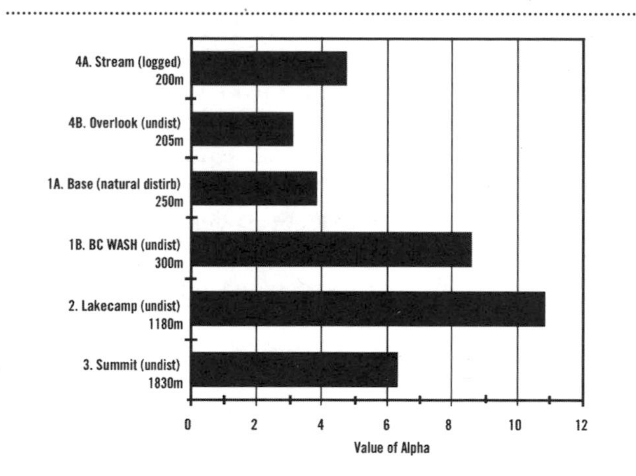

Figure 3.8. Alpha values for green geometrid moths samples at the six sites.

Effects of Collecting Intensity

As expected, there was no association between either (a) the number of hours spent collecting moths at a site, or (b) the number of total moths collected, and (c) the *alpha index*. At the same time, small samples typically lead to extremely high *Chi-square* values in these calculations, and *alpha* values for each night's samples can vary tremendously. Thus, sample size and collecting intensity *did* affect species richness calculations, and the data were used to investigate this further.

Figure 3.9 shows for the Lake Camp (Site 2) that as the data from consecutive nights were summed, that *alpha* increased, in a nearly linear manner, for ten nights, with no strong indication that this increase would halt with additional collecting. Figure 3.10 shows that the accumulation of species at the Lake Camp closely paralleled that for *alpha*, again with no indication that ten days of sampling had fully sampled the moth fauna. This result has two disenchanting interpretations. First, the moth fauna in Melanesia may, in fact, be too species-rich to sample in a short period appropriate to rapid survey. A second interpretation might be that non-area-based "attraction" lighting of moths is not appropriate to site-based sampling because (a) there is no way to determine the "sampling radius" of the moths being captured over time, and (b) such an attraction-type sampling method may not be expected to reach a species-accumulation asymptote in short-term efforts.

The Commercial Value of Southern New Ireland Insects

Forty-six species of butterflies were recorded during the survey, probably only slightly more than half of the butterfly

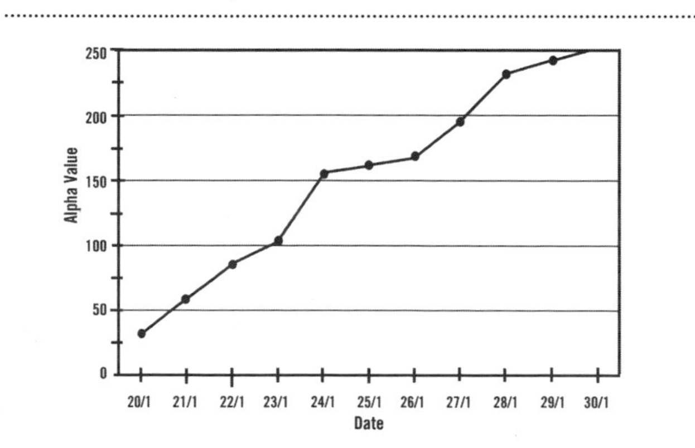

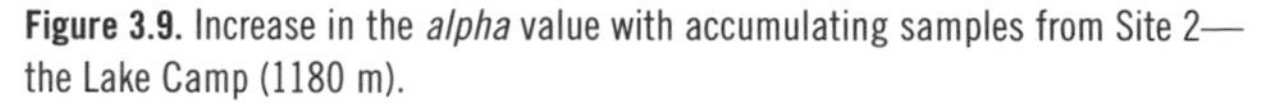
Figure 3.9. Increase in the *alpha* value with accumulating samples from Site 2—the Lake Camp (1180 m).

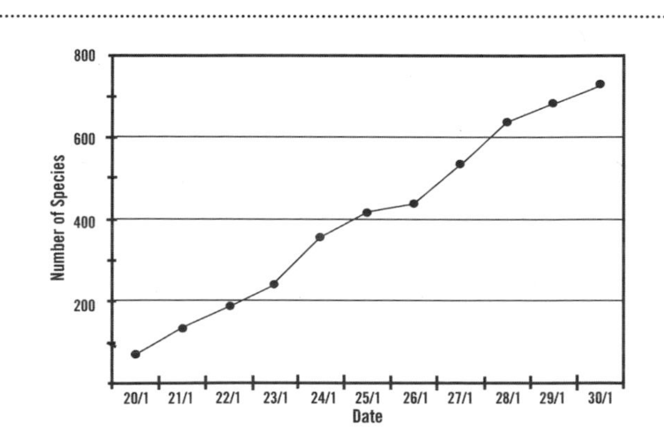

Figure 3.10. Moth species accumulation over time from Site 2 at the Lake Camp station. Note that no asymptote is reached even after 11 days of sampling.

fauna for the area (Appendix 2). The discovery of one new species of butterfly, *Graphium kosii* (Muller and Tennent 1999) of very localized distribution, gives the area potential for local income generation from collected insects. At the same time, the most valuable species are probably confined to the inaccessible interior highlands. A hike of at least two days is required to reach the 2000+ meter elevation where the new *Graphium* swallowtail is common. Although the beetles and walkingsticks collected would sell for relatively little on the commercial market they are much more accessible. The only easily farmable butterfly found near human habitation was the birdwing, *Ornithoptera priamus urvilliana.* Several farms could be established in the coastal strand. Because 22 of the 46 recorded butterflies are taxa restricted to New Ireland or the Bismarck Archipelago, specimens of these geographically restricted taxa command a slight premium over the more widespread mainland relatives.

Discussion

Effects of Elevation on Moth Species Richness

The data indicate a trend that has been identified for other plant and animal groups, both on New Guinea and elsewhere in the tropics; that is, increasing species richness to foothill elevations, then a decline with increasing elevation. Gressitt and Nadkarni (1978) felt that the highest beetle diversity occurred at roughly 1100-1200 meters in the Wau region of mainland PNG; Holloway (1987) found the highest *alpha* values for Borneo moths at around 1000 meters. Elevations anywhere from about 400 meters to 1200 meters are mentioned as the optimal elevational range for species diversity in the tropics. Thus it is apparent that our expedition failed to establish a field camp at the elevational zone that is probably the richest on the island.

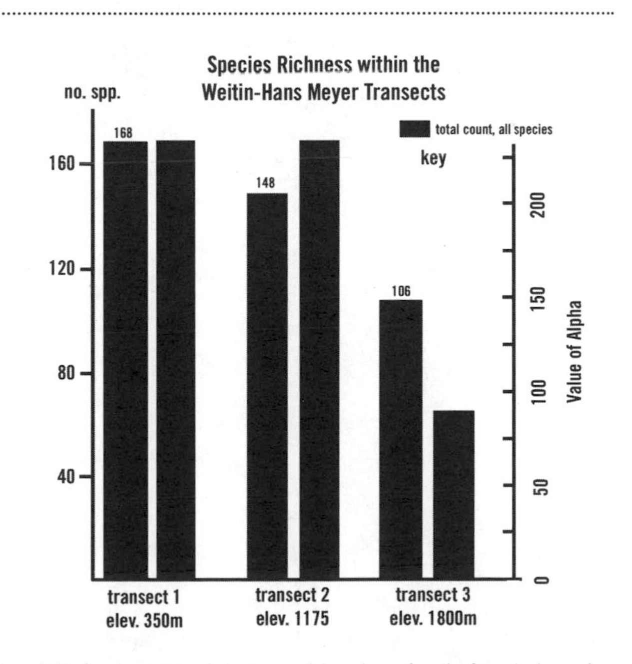

Figure 3.11. Species counts of plants vs. *alpha* values of moths for paired samples in the Weitin watershed.

Diversity of Moths and Plants

Moth collecting sites at the Weitlin Base Camp (1B), Lake Camp (2) and High (Top) Camp (3) were near the plant enumeration transects established by Wayne Takeuchi and Joseph Wiakabu on this expedition, providing a special opportunity to compare the species richness of the two groups for different sites. The plant results (the number of species enumerated in the standard transects) and the moth collecting results (shown as *alpha indices*) are shown in Figure 3.11. Because only three pairwise comparisons are possible, no statistical test of correlation is merited. In any event, the lack of close correspondence indicates either that there is no tracking of plant diversity by moth diversity or else that more data are needed to delineate such a pattern.

Effects of Logging and Disturbance on Moth Species Richness

Holloway's (1987) survey of moth diversity in disturbed and undisturbed forest in Sulawesi showed that the disturbed forests supported fewer moth species. Likewise, the two disturbed sites in the New Ireland transect had lower *alpha* values than their undisturbed counterpart sites. The commercially logged stream site showed the lowest species richness of any site, apart from the High Camp (1830 m).

Moth Species Diversity on New Ireland

The *alpha* values calculated at the southern New Ireland sites are much higher than those obtained from lower montane Hunstein Mountain samples (Balun et al, 1994), even though the family composition of the Hunstein sample was comparable to sites at equivalent elevations in southern New Ireland. However, in general, higher diversity would be expected on the mainland, with some evidence on moths to support this. The New Ireland results show less diversity overall than for larger land masses like Borneo and mainland New Guinea, which is expected. But the New Ireland results indicate higher diversity than for Sulawesi, which is somewhat unpredicted, since Sulawesi is considerably larger than New Ireland. New Ireland, being closer to a large land mass (New Guinea), may "capture" a greater percentage of the mainland community as a result.

The Lost *Papilio moerneri* Swallowtail Butterfly

We failed to find *Papilio moerneri*, known from only a few specimens from New Ireland. *P. moerneri* is closely related to a reasonably common Bougainville species (*Papilio toberi*) and a common mainland species (*Papilio laglezei*). As the life histories of those two species are similar, we believe

P. moerneri's behaviour and biology can be predicted on the basis of the traits of its outlying relatives. Future workers should focus on (1) searching for the caterpillars on Lauraceaous tree species (especially the genus *Litsea*); (2) searching for gregarious assemblages of the caterpillars; and (3) searching for a caterpillar that is likely to be black-striped longitudinally, alternating with perhaps orange or whitish bands.

Conclusions

1. Southern New Ireland's insects include a number of commercially exploitable species, some of value on the world market. The birdwing butterfly *Ornithoptera priamus urvilliana,* locally common and easily farmed, offers perhaps the best opportunity for sustainable production and sale.

2. Distinctive moth faunas occupy the higher elevation cloud forests, and to a lesser extent, mid-elevation forests (1180 meters). These high elevation moth communities are characterised by "site-specific" species, many of which may prove to be endemic to montane New Ireland. The presence of the distinct *Graphium kosii* swallowtail further supports the uniqueness of the higher forest habitats. There also appears to be a distinctive, coastal community that was not rigorously sampled.

3. Intensive commercial logging seems to have caused an impoverishment of the moth community. This species impoverishment seen for moths may also occur in plants and perhaps other animal groups. By decreasing the diversity of plants and animals, logging has reduced the future options and the future value of those forests to the customary landowners.

4. Green geometrine moths, previously suspected to be possible "flagships" of exceptionally species-rich areas in New Guinea, instead appear to be indicators of high rainfall sites. As such, they also may be good indicators of sites with a high epiphyte plant component.

5. Night-to-night fluctuations in *alpha* are often considerable, making single-night collections unreliable for comparing sites. Again the problem is with the low sample size associated with a single night of collecting.

6. Undisturbed coastal forests remain inadequately surveyed for plants and animals. The moth survey indicates that this may be a community with members that are not found in the interior Weiten valley. Any sustainable logging operation should operate in conjunction with set-aside reserve areas at these lower elevations, unless follow-up wildlife surveys indicate that this is unnecessary.

Acknowledgments

Foremost, we would like to thank the customary landowners of southern New Ireland for permission to carry out the survey on their land. We thank our fellow RAP Team members and our field assistants for their help and companionship during the survey. We are most grateful to staff members of the PNG Department of Environment and Conservation, particularly Samuel Antiko and Felix Kinbag, who were responsible for many logistical arrangements. Bruce Jefferies is thanked for his overall facilitation and encouragement. Finally, this study could not have been possible without the diligent work of Andrew Kinibel, Christopher Dal, Joseph Samp, Barry Andreas, and Hais Wasel. We are grateful to the *The Christensen Fund* (Palo Alto, California), which provided facilities and staff time for the drafting of this report.

Literature Cited

Balun, L., A. Vitas, and L. Orsak. 1994. An assessment of the biological diversity of the Hunstein Range, East Sepik Province, Papua New Guinea. (Unpublished). *Christensen Research Institute*, Madang, Papua New Guinea. 46 pp.

Fisher, R.A. 1943. A theoretical distribution for the apparent abundance of different species, Part 3, pp. 54–57. *In:* Fischer, R.A., A.S. Corber and C.B. Williams (eds.), The relation between the number of species and the number of individuals in a random sample of an animal population. *Journal of Animal Ecology* 12:42–58.

Gressitt, J.L., and N. Nadkarni. 1978. Guide to Mt. Kaindi: background to montane New Guinea ecology. Wau Ecology Institute Handbook #5. 135 pp.

Hebert, P.D.N. 1980. Moth communities in montane Papua New Guinea. *Journal of Animal Ecology* 49: 593–602.

Holloway, J. 1976. The moths of Borneo, with special reference to Mt. Kinabalu. Malayan Nature Society, Kuala Lumpur. 264 pp.

Holloway, J. 1984. Moths as indicator organisms for categorising rain-forest and monitoring changes and regeneration processes. *In: Tropical Rain-Forest: The Leeds Symposium*, eds. A.C. Chadwick and S.L. Sutton. Leeds, United Kingdom. Leeds Philosophic and Literary Society.

Holloway, J. 1987. Macrolepidoptera diversity in the Indo-Australian tropics: geographic, biotopic and taxonomic variations. *Biological Journal of the Linnean Society* 30:325–341.

Kempton, R.A. 1975. A generalised form of Fisher's logarithmic series. *Biometrika* 62: 29–38.

Kempton, R.A., and L.R. Taylor. 1974. Log-series and log-normal parameters as diversity discriminators for the Lepidoptera. *Journal of Animal Ecology* 43: 381–399.

Magurran, A.E. 1988. Ecological diversity and its measurement. Princeton University Press, Princeton, New Jersey. 125 pp.

Muller, C.J. and W.J. Tennent. 1999. A new species of *Graphium* Scopoli (Lepidoptera: Papilionidae) from the Bismarck Archipelago, Papua New Guinea. Records of the Australian Museum 51: 161–168.

Southwood, T.R.E., V.K. Brown, and P.M. Reader. 1979. The relationship of plant and insect diversities in succession. *Biological Journal of the Linnean Society* 12:327–348.

Taylor, L.R, R.A. Kempton and I.P. Woiwod. 1976. Diversity statistics and the log-series model. *Journal of Animal Ecology* 45:255–272.

Thomas, A.W. and G.M. Thomas. 1994. Sampling strategies for estimating moth species diversity using a light trap in a northeastern softwood forest. *Journal of the Lepidopterists Society* 48(2):85–105.

Chapter 4

The Herpetofauna of Southern New Ireland

Allen Allison and Ilaiah Bigilale

Abstract

During 32 days of field work at three main sites and a number of additional collecting localities in the Weitin Valley, southern New Ireland, we collected 351 specimens of amphibians and reptiles, representing 39 species. This total included seven species of frogs, 20 lizards and 12 snakes, and comprises 79% of the 49 amphibian and reptile species now known to occur in New Ireland. One of the frogs was a new species now described as *Platymantis browni,* and one or more of the lizards may also be undescribed. Three species are reported for the first time from New Ireland, including a range extension for *Typhlops depressiceps*, a blind snake previously known only from the mainland of New Guinea but more recently documented from New Britain. We briefly report on the abundance of each species, its habitat requirements, and known geographic distribution. Our study confirms that southern New Ireland is a center of high species diversity for the island.

Introduction

New Ireland is situated just north and east of New Britain in the Bismarck Archipelago, between the islands of Buka and Bougainville to the southeast and the Admiralty Archipelago to the west (Map 1). This region is geologically complex and is "a maze of small oceanic basins and active and extinct island arcs" (Hamilton 1979), along which amphibians and reptiles have dispersed from source areas in the Indo-Australian region, and undergone independent radiations, to inhabit the islands of the Bismarck and Solomon Seas. The present-day amphibians and reptiles of the Bismarck Archipelago reflect this complex history and comprise a mixture of widespread Indo-Pacific species that have arrived in geologically recent times (some probably aided by humans) and regional and island endemics that have much older local histories.

The herpetofauna of New Ireland is poorly known, but on the basis of scattered museum records is known to be similar to that of New Britain (Werner 1898, 1900, Hediger 1934, Mys 1988). Most of the species previously recorded from New Ireland are also known from New Britain, including two species of colubrid snakes, *Stegonotus heterurus* and *Tropidonophis hypomelas*, and one species of skink, *Emoia bismarckensis*, all endemic to the Bismarck Archipelago. However, a number of genera known from surrounding regions have not been recorded from New Ireland, including *Tribolonotus* (Scincidae) (Cogger 1972, Zweifel 1966), *Discodeles* (Ranidae)(Brown and Webster 1969), and *Oreophryne* and *Sphenophryne* (Microhylidae) (Tyler 1964, 1967). Many of the species in these and other genera inhabiting the Bismarck Archipelago are easily overlooked by general collectors and have only recently been discovered and described. This suggests that the small size of the known New Ireland herpetofauna, and the absence of expected genera, may simply reflect inadequate collecting and our poor knowledge of the island (Brown 1983).

The Papua New Guinea Conservation Needs Assessment (Beehler 1993) strongly recommended additional field study in New Ireland. Two areas were singled out as important centers of biodiversity: the Lelet Plateau in the center of the island and the mountainous region in the far south (Hans Meyer and Verron Ranges). The southern region is the most poorly known part of New Ireland (prior to the expedition, almost no amphibians and reptiles had been recorded from there) but, on the basis of high topographic diversity, was postulated by Allison (1993) to be an important center of herpetological diversity on the island.

Previous Work on the New Ireland Herpetofauna

Detailed knowledge of the herpetofauna of the Bismarck Archipelago dates from the mid-to-late-1800s when German missionaries on the Duke of York Islands began depositing collections in European museums (Günther 1877). By the turn of the century, a number of collectors had also visited adjacent islands and the existing knowledge of the herpetofauna of the Bismarck Archipelago was summarized by Werner (1900), who recorded a total of 43 species for the region, but did not specifically list any documented records from New Ireland. Collections continued to accumulate in association with German business activity in the Bismarck Archipelago.

In 1934 Hediger, who spent an extended period in the region, summarized what was then known, listing 27 species of reptiles from New Ireland. He recorded for the first time a number of species that are now known to be abundant throughout much of the SW Pacific region (e.g., *Nactus pelagicus*) and indicated that all but two of the species that he listed were known also from New Guinea or the Solomon Islands. The exceptions were *Lipinia rouxi,* which he described as new, and *Acutotyphlops subocularis,* which was described as *Typhlops subocularis* by Waite (1897) from the Duke of York Islands. *Lipinia rouxi* is now known to be endemic to New Ireland, and *A. subocularis* is endemic to the Bismarck and Solomon Archipelagoes. Hediger (1934) also made extensive ecological observations and commented on the zoogeography of the Bismarck herpetofauna.

A number of collectors have since visited the Archipelago, and their collections have generally been reported in monographic studies of various taxa (e.g., Zweifel 1960, Tyler 1968, McDowell 1967, 1972, 1974, 1975, 1979, Malnate and Underwood 1988, Kluge 1993). McCoy (1980) provided a comprehensive treatment of the reptiles of the Solomon Islands; many of those species occur also in the Bismarck Archipelago.

Methods

We conducted a field survey of amphibians and reptiles with our four objectives being as follows: (a) the compilation of a complete species inventory of the terrestrial and freshwater herpetofauna, (b) the discovery of undescribed species or species not previously reported from New Ireland, (c) a simple assessment of the abundance of each species and their general habitat preferences, and (d) an assessment of threats to the survival of any particularly rare or noteworthy species.

Itinerary

We arrived at the Weitin Base Camp (ca. 240 m elevation) along the Weitin River on 14 January 1994 and remained there for 11 days (see Map 2). On 24 January, we traveled by helicopter to a camp adjacent to a small lake at ca. 1200 m in the Hans Meyer Range (Lake Camp) and remained there for 9 days. On 1 February, Allison hiked to a third camp at ca. 1800 m (Top Camp) and remained there for two days. Bigilale remained at the Lake Camp during that time. On 3 February, we returned by helicopter to a lowland site on the floor of the Weitin Valley along a small tributary to the Weitin River (Riverside Camp, ca. 150 m) where we stayed until leaving New Ireland on 16 February.

Collecting Protocol

Together with two full-time assistants, each day we collected amphibians and reptiles by hand around each of the camps from early morning until late in the night. In addition, we arranged for local people to collect in coastal areas on either side of the mouth of the Weitin River. To assist in this effort we handed out nearly 1000 large rubber bands for use in stunning small, fast-moving lizards.

We searched in all appropriate habitats, turning over logs and stones, peeling bark from dead trees, using headlamps to locate nocturnal species (many species have a distinctive eye shine). Frogs were also located by call. In addition, while based at the Riverside Camp, we drove at night on several occasions along logging tracks and the coastal road looking for amphibians and reptiles. This method is particularly effective for finding snakes.

The calls of at least one individual of each frog species were recorded (Sony WCMD6 with Sony ECM 270F microphone), and nearly all species of amphibians and reptiles were photographed alive.

All specimens selected as vouchers were killed with injections of Nembutal, fixed for several days in 10% buffered formalin and then stored in 70% ethanol. Prior to preservation, liver tissue was removed from a few specimens and preserved for biochemical study. Most specimens were measured when freshly killed. For frogs, only the distance from the tip of the snout to the vent (SVL) was measured, but for lizards and snakes the total length (TBL—tip of snout to tip of tail) as well as SVL were measured. These details were recorded, together with observations on color and other features.

Specimens have been deposited in the Papua New Guinea National Museum, Port Moresby, and the Bernice P. Bishop Museum, Honolulu, Hawaii.

Results

We preserved a total of 351 specimens, including 208 frogs, 119 lizards, and 24 snakes, comprising 39 species (see list below). No native species were seen at the 1800 m camp, either by Allison during a two-day visit or by other members of the survey who spent more than a week at that site. A single individual of the introduced species, *Bufo marinus*, was sighted by Louise Emmons at 2200 m (see below).

The terrestrial and freshwater amphibians and reptiles now known from or likely to occur in southern New Ireland include 49 taxa: 8 frogs, 1 crocodile, 27 lizards and 13 snakes. The 45 taxa that can be identified to species can be divided into six categories on the basis of their geographic distribution (Table 4.1). Approximately two-thirds of New Ireland amphibian and reptile taxa are also found in New Guinea and the Indo-Australian Archipelago. The remaining third are endemic to the islands of the Bismarck and Solomon Seas, with about half of these being endemic to the Bismarck Archipelago and the other half shared with the Solomon Islands.

Table 4.1. General distribution of New Ireland amphibians and reptiles species by geographic range. Categories are mutually exclusive and do not include two species, *Cryptoblepharus poecilopleurus* and *Eugongylus rufescens*, that in the New Ireland region are known only from offshore islands to the north of New Ireland.

Geographic Range	# Species
Introduced	1
Widespread—Indo-Pacific	20
Widespread—Papuan Region	15
Bismarck and Solomon Island Archipelagoes	4
Endemic to Bismarck Archipelago	7
Presumed Endemic to New Ireland	2
Total	49

In the list that follows, the number of specimens we collected is shown in parentheses after the species name; for described species the most current or comprehensive references are cited; species names follow Allison (1993, 1996 updated to include newly described species). In this treatment, lowland refers to the Weitin Base Camp (240 m) and all areas below this, including the Riverside Camp, the logging road to the coast, and all coastal areas. The Weitin Valley refers to all areas up to ca. 1200 m elevation.

Details on the collection, including specimen catalog numbers, will be made available via the Bishop Museum's Web site. Some background details on the collection sites are available in Allison and Kraus (2001).

Amphibians and reptiles of southern New Ireland

Amphibians: Frogs

Bufonidae

Bufo marinus (1)—Abundant throughout the Weitin Valley. Individuals ranging in size from juveniles to large adults were active mostly at night in primary forest, regrowth areas, and open river beds. One very large individual was found resting on a bed of deep moss by Louise Emmons at 2200 m in the Hans Meyer Range, the only amphibian or reptile recorded above 1800 m. This toad is native to tropical America and was introduced to New Guinea. Bailey (1976) reported on its diet in a cacao plantation in New Britain and Zug *et al.* (1975) provided details on its distribution and ecology in Papua New Guinea.

Hylidae

Litoria infrafrenata (2)—Abundant at the Lake Camp. Males called throughout the day and night from within the low vegetation surrounding the lake. This species was also heard calling in abundance from swamps along the coast near the mouth of the Weitin River. We collected both of our specimens as they crossed a logging road at night in lowland forest near the coast. This species occurs throughout New Guinea and the Bismarck Archipelago and was first reported from New Ireland by Ramsay (1878), who described it as *Pelodryas militarius.* This is now regarded as *Litoria infrafrenata militaria,* confined to the Bismarck Archipelago (Tyler 1968).

Litoria thesaurensis (30)—Abundant throughout the Weitin Valley. Most individuals were perched in small trees or shrubs at night, mostly within 1–2 m above the ground in closed canopy forest. There were also large breeding aggregations in the lake at the Lake Camp and also in small forest ponds near the Riverside Camp. Many of the females we collected were gravid. *Litoria thesaurensis* has been reported from much of New Guinea as well as the Solomon Islands, Bougainville, and New Hanover by Tyler (1968), who speculated that it undoubtedly occurred on New Ireland. Brown and Menzies (1978) list it, without comment, as occurring on New Ireland.

Ranidae

Platymantis browni (61)—Abundant throughout the Weitin Valley to ca. 1350 m. The activity patterns and habits of this undescribed species were similar to those of *P. schmidti,* except that *Platymantis browni* called from perches in the forest understory that were generally ca. 1 m or less above the forest floor. It was the smallest species of *Platymantis* occurring in the Weitin Valley (adult SVL 20–25 mm) and may have been overlooked by previous workers who assumed that it was the young of one of its congeners occurring in the area. It is now known to also occur in New Britain (observations by Stephen Richards reported in Allison and Kraus 2000).

Platymantis magnus (6)—Rare at the Lake Camp but locally common at the Riverside Camp. A frog that we presume to be this species gave several calls beginning at 2300 hr on 23 January following a heavy rain at the Lake Camp. We did not see or hear any other individuals during the week we spent at that camp. Numerous individuals were heard calling around the Riverside Camp (150 m) following heavy rains. This species, which is strictly terrestrial, seemed to prefer thick underbrush adjacent to streams and gullies and called only after heavy rain, beginning 1–2 hours after nightfall. *P. magnus* was described by Brown and Menzies (1978) from northern New Ireland and is currently known only from that island.

Platymantis papuensis (63)—Abundant on the forest floor to at least 1400 m elevation. Males produced loud, chattering calls intermittently during the day from hidden retreats and, with far greater frequency, from exposed locations throughout much of the early evening. Calling frequency generally tapered off gradually as the night advanced. Calling activity increased noticeably after late afternoon or early evening rains but even on dry nights a few individuals called. This species has been reported from New Ireland by Menzies (1982a; 1982b). It occurs also on the north coast of New Guinea.

Platymantis schmidti (45)—Abundant in low understory vegetation throughout the Weitin Valley. Males called from perches in small trees or shrubs ca. 2 m above the ground. Most individuals began calling just after dark if there had been recent rain. *P. schmidti* is apparently endemic to New Britain and New Ireland (Brown and Tyler 1968, Menzies 1982a), although Menzies (1982a) also lists a questionable record from Manus Island.

Rana kreffti—Not observed during our survey. We searched suitable habitats for this species, which frequents lowland swamps, but did not see or hear it, although it was reported from New Ireland by Menzies (1987) who listed its range as "the entire Solomon Island chain from San Cristobal in the south to Buka in the north; New Ireland."

Reptiles: Crocodiles

Crocodylidae

Crocodylus porosus—Not observed during our study. Local villagers reported that this large crocodile was common in coastal rivers and swamps in the Weitin Valley. The species is widely distributed from Asia to the Solomon Islands (King and Burke 1989) and was reported by Werner (1900) and Hediger (1934) from the Bismarck Archipelago.

Reptiles: Lizards

Agamidae

Hypsilurus godeffroyi—Not observed during our survey. Wermuth (1967) listed this species as occurring in the Admiralty, Bismarck, and Solomon Archipelagoes and Hediger (1934) tentatively listed it as occurring in New Ireland. In his remarks, Hediger expressed uncertainty whether his specimens were from the Duke of York Islands or the adjacent coast of New Ireland. *Hypsilurus godeffroyi* is known to be highly arboreal and is easily missed (Bauer and Watkins-Colwell 2001), although we did make a special effort to find it.

Hypsilurus modestus (3)—Common in small trees along the edge of closed canopy forest throughout the Weitin Valley. As with most agamids in the New Guinea region, this species was highly arboreal. It is known from lowland regions throughout mainland New Guinea and the Bismarck and Admiralty Archipelagoes (Wermuth 1967).

Gekkonidae

Gehyra mutilata (1)—The only individual that we observed was captured in a village house on the coast near Silur. This common gecko is widely distributed throughout the Indo-Pacific region and has been reported from New Ireland by Bauer (1994).

Gehyra oceanica (8)—Probably common, although the only individuals that we saw were taken together in an abandoned village house along the coast. All attempted to bite and to twist violently when grabbed, often losing large portions of their skin and their tails. *Gehyra oceanica* has been reported previously from New Ireland; it also occurs on New Britain, the Moluccas, New Guinea and throughout much of the Pacific Basin (Bauer 1994).

Gekko vittatus (4)—Common along the coast in village houses. None was seen in areas around the field camps. This distinctive gecko is widely-distributed from Java to New Guinea and throughout the Admiralty, Bismarck, and Solomon Archipelagoes (Bauer 1994).

Hemidactylus frenatus (4)—Common in the buildings of a logging camp near the mouth of the Weitin River. None was seen in the forest around any of the field camps. *Hemidactylus frenatus* is native to SE Asia and the Indo-Australian Archipelago. It is a human commensal and is now found in tropical and subtropical regions throughout the world (Bauer 1994).

Lepidodactylus lugubris—Not collected during this survey. This species is widely distributed throughout the Pacific Basin and was listed from New Ireland by Bauer (1994).

Nactus sp. (*pelagicus* group) (9)—Abundant in the forests around all the lowland field camps in the Weitin River Valley. All individuals were strictly nocturnal and were generally perched on the buttresses of large trees in closed canopy forest 1–2 m above the ground, and were almost always solitary. This species has generally been reported in the literature as *Nactus pelagicus*. However, that species is a complex of sexual and asexual populations, many of which probably represent valid species. This problem is currently under study by George Zug and colleagues at the U.S. National Museum of Natural History and is beyond the scope of this paper. The nominal species has been reported from New Ireland and is widely distributed from the Kei Islands through New Guinea to the islands of the Pacific Basin (Bauer 1994).

Scincidae

Carlia sp. (probably an undescribed species in the *fusca* group) (20)—Abundant in leaf litter throughout the Weitin Valley. Most individuals were observed in open riverine or disturbed forest. This species was absent from the Lake Camp, suggesting that it is restricted to lowland habitats.

The *Carlia fusca* superspecies is known from islands throughout the SW Pacific, including New Ireland (Mys 1988).

Cryptoblepharus poecilopleurus—Not collected during our survey. Mys (1988) lists the species from the islands off the north coast of New Ireland; there are apparently no confirmed records of it occurring on mainland New Ireland. *Cryptoblepharus poecilopleurus* is widespread among the islands of Oceania (Allison 1996).

Emoia atrocostata (2)—Abundant in coastal areas. All individuals were seen basking on rocks in the intertidal zone. This skink occurs throughout the Indo-Australian Archipelago and the SW Pacific and was reported from the Bismarck Archipelago by Brown (1991).

Emoia bismarckensis (2)—Rare in leaf litter in open riverine or disturbed forest. The only representatives that we saw were around the lowland camps. Fewer than 10 specimens have previously been collected from New Ireland. It is endemic to the Bismarck Archipelago (Brown 1991).

Emoia caeruleocauda (12)—Abundant in leaf litter in open areas in primary and second-growth forest at the Lake Camp and below. This colorful skink occurs from the southern Philippines through the Palau, Caroline and Marshall islands and from Borneo through the Celebes, New Guinea, to the Bismarck and Solomon Archipelagoes (Brown 1991).

Emoia cyanogaster (3)—Common in low vegetation in open, disturbed areas such as gardens, cacao plantations, etc., along the coast. Most individuals were observed climbing in low vegetation. Brown (1991) reported on specimens from New Ireland and listed its range as the Bismarck and Solomon Archipelagoes and Vanuatu.

Emoia impar (5)—Common in leaf litter in open riverine or disturbed forests around the Lake Camp and below. This species is difficult to distinguish at a distance from sympatric *E. caeruleocauda,* which it superficially resembles. *Emoia impar* has long been relegated to synonomy under *E. cyanura* (Brown 1991, Ineich and Zug 1991). However, *E. cyanura* is now known to consist of a complex of at least two closely related species (the other being *E. impar*), that occur in broad sympatry throughout the islands of Oceania (Ineich and Zug 1991; Bruna et al. 1995). *E. impar* was the only species in this complex that we recorded from the Weitin Valley. Although *E. cyanura* is recorded from New Ireland (Hediger 1934; Mys 1988), this may refer to *E. impar*. Mys (1988) mentioned that the *E. cyanura* he collected on the Lelet Plateau seemed different from typical *E. cyanura.* We have not yet reexamined material in other collections and can at this point say with certainty only that *E. impar* is present in New Ireland. Both *E. cyanura* and *E. impar* are present on New Britain (Ineich and Zug, 1991).

Emoia jakati (17)—Abundant in leaf litter in open riverine or disturbed forests in the lowlands of the Weitin Valley. This was probably the most common lizard in the area. Often considered a reptilian weed, *E. jakati* occurs from the Palau, Marshall, and Caroline Islands to lowland New Guinea and the Bismarck and Solomon Archipelagoes (Brown 1991).

Eugongylus albofasciolatus (1)—Apparently rare or uncommon in the Weitin Valley. The single individual we observed was collected under a pile of rotting vegetation on the forest floor around the Riverside Camp. This species is found from the Moluccas, through New Guinea and the northern Cape York Peninsula (record uncertain—see Cogger [1992]), to the Bismarck and Solomon Archipelagoes (McCoy 1980).

Eugongylus rufescens—Not collected during our survey. Mys (1988) lists it from the islands off the north coast of New Ireland; there are apparently no confirmed records of it occurring on mainland New Ireland. It occurs from the Moluccas and Lesser Sundas across New Guinea and the Cape York Peninsula of Australia, through the Admiralty and Bismarck archipelagos to the Solomon Islands (Rennell only) (McCoy 1980).

Lamprolepis smaragdina (6)—Abundant in open riverine or disturbed forests at lowland sites throughout the Weitin Valley. This distinctive, fast-moving green skink was especially abundant along the coast in copra plantations. It is highly arboreal and generally occurred on tree trunks. Hediger (1934) listed it from New Ireland. The species is widely distributed on major islands throughout SE Asia to New Guinea, northern Australia, Micronesia and the Solomon Islands (Greer 1970).

Lipinia noctua—Not collected during our survey. Greer and Mys (1987) listed this species as being sympatric in New Ireland with *L. rouxi. Lipinia noctua* ranges from the Moluccas, New Guinea to the Solomon Islands and much of tropical Oceania (McCoy 1980).

Lipinia rouxi—Not collected during our survey despite intense searches of appropriate habitat. This species, described by Hediger (1934) from New Ireland, was placed in synonymy with *Lipinia noctua* by Zweifel (1979) and resurrected by Greer and Mys (1987) who showed that it was anatomically distinct from *L. noctua. Lipinia rouxi* is endemic to New Ireland (Greer and Mys 1987).

Sphenomorphus solomonis—Not collected during our survey. The species was reported from New Ireland by Hediger (1934) and Greer and Parker (1974). The latter authors listed the species range as Halmahera and Ternate, east through northern New Guinea, and the Admiralty and Bismarck Archipelagoes to the eastern Solomon Islands.

Sphenomorphus simus (13)—Abundant in lowland sites throughout the Weitin Valley. Most individuals were found under logs, piles of vegetation, etc. Hediger (1934) listed it (as *S. variegatus*) from New Ireland. *Sphenomorphus simus* ranges from Halmahera and the Moluccas east to New Guinea and the Bismarck Archipelago (Mys 1988).

Sphenomorphus tanneri—Not collected during our survey; this species is known from New Britain, New Ireland, Bougainville, Choiseul and the Shortland Islands (Mys 1988).

Sphenomorphus cf. *jobiensis* (6)—Abundant in closed canopy forest in the lowlands of the Weitin Valley. All individuals that we observed were strictly terrestrial and apparently crepuscular. The nominate species was reported from New Ireland by Hediger (1934), but *S. jobiensis* is now thought to comprise a species complex (Donnellan and Applin 1989) of which there are at least two forms in New Ireland. A detailed taxonomic analysis is beyond the scope of this paper, so we are referring to the taxon most similar to the nominate species as *S.* cf. *jobiensis* and the other as *Sphenomorphus* sp. (see below). The *S. jobiensis* species complex occurs throughout New Guinea and nearby islands (Mys 1988).

Sphenomorphus wolfi (1)—One individual was collected from under a box at the Lake Camp; another that was seen nearby escaped. This is a secretive species that shelters under fallen logs and other debris on the forest floor. This taxon has often been misidentified in the literature as *S. maindroni*, but that species is now thought to be restricted to the New Guinea mainland (see discussion in Mys 1980). *Sphenomorphus wolfi* has also been confused with *S. derooyae* (Mys 1988). According to Glenn Shea of Sydney University, who identified our specimens, *Sphenomorphus wolfi* is restricted to the Admiralty and Bismarck archipelagos and *Sphenomorphus derooyae* is restricted to the north coast lowlands of New Guinea.

Sphenomorphus sp. (1)—A large species in the *S. jobiensis* complex (SVL = 113 mm). We collected the sole representative from leaf litter in closed canopy forest around the Riverside Camp (150 m). This taxon may be referable to *S. megaspilus* (Günther 1877), but this requires further study (see comments on *S. jobiensis*, above).

Varanidae

Varanus indicus (1)—The only individual we saw was collected from coastal forest near Silur. Local villagers reported that this species was common in the area. In other parts of its range this species is both arboreal and terrestrial, moving quickly through the forest floor, fleeing long distances when disturbed. It is listed from the Bismarck Archipelago by Hediger (1934) and is known to occur throughout much of the Indo-Australian Archipelago and islands of the SW Pacific.

Varanus finschi—Not collected during our survey. This recently described species in the *indicus* group is now known to occur in Australia, New Britain, and New Ireland (Philipp et al. 1999).

Reptiles: Snakes

Boidae

Candoia aspera (1)—The only individual we observed was collected from the forest floor during the day in second growth vegetation near the mouth of the Weitin River. Local villagers reported that this species was common in the Weitin Valley and adjacent coastal areas. This distinctive boa ranges throughout much of New Guinea and nearby islands to the Bismarck and Admiralty Archipelagoes (McDowell 1979).

Candoia carinata (2)—Apparently common in lowland areas, according to local villagers. The only individuals that we saw were both collected during the day from the forest floor in second growth areas adjacent to villages near Silur. McDowell (1979) reported this species from New Ireland and gave its range as Palau, Celebes, and Moluccas east through New Guinea to the Admiralty, Bismarck, and Solomon Archipelagoes.

Pythonidae

Bothrochilus boa (2)—Apparently common in lowland areas. Individuals were observed only at dusk or at night, along logging roads through closed canopy forest, or along roads in open coastal forest interspersed with copra and cacao plantations or, in three instances, in dry rocky/sandy areas near closed canopy forest fringing river drainages. One individual regurgitated a *Dendrelaphis* suggesting that *B. boa* is, at least in part, a snake predator. This python is endemic to the Bismarck Archipelago and Bougainville (Kluge 1993).

Morelia amethistina (2)—Common in the lowlands of the Weitin Valley. This species was regularly observed along logging roads through heavily disturbed forest and was strictly nocturnal. It occurs from the Moluccas, Timor, and Halmahera throughout New Guinea and the Cape York Peninsula to the Bismarck Archipelago (McDowell 1975).

Colubridae

Boiga irregularis (4)—Common at all lowland sites. This species may occur to at least 1500 m (an individual thought to be this species was observed by Bruce Beehler at that elevation). Almost all individuals were observed climbing on shrubs or small trees in the understory of closed canopy forest and along logging roads. It is strictly nocturnal. This species occurs throughout New Guinea and nearby islands, northern and eastern Australia, and the Admiralty, Bismarck, and Solomon Archipelagoes (de Rooij 1917, McCoy 1980). It was first reported from the Bismarck Archipelago by Hediger (1934).

Dendrelaphis punctulata (3)—The *Dendrelaphis* in New Ireland is apparently referable to this species (O'Shea 1996), although our samples seemed more similar to *D. calligastra*. Our taxon was common at all lowland sites where it was seen or collected during the day in both primary forest and heavily disturbed or cultivated areas. Individuals were terrestrial and arboreal, and were often found in small trees and shrubs. This species was listed from New Ireland by Hediger (1934) as *Dendrophis lineolatus* which is now considered a subspecies of *punctulata. Dendrelaphis punctulata* occurs throughout New Guinea and associated offshore islands, the northern and eastern coasts of Australia and the Bismarck Archipelago (de Rooij 1917, Cogger 1992). The closely related species, *Dendrelaphis calligastra,* is known from Indonesian islands in the Arafura Sea, throughout New Guinea, northern Australia, and the Solomon Islands (de Rooij 1917, McCoy 1980) and has been reported from the Bismarck Archipelago by Hediger (1934).

Stegonotus heterurus (3)—Seen regularly at night on the floor of closed canopy forest around the Weitin Base and Riverside Camps, and is probably common. This snake was reported from New Ireland by McDowell (1972), who listed it as also occurring on New Britain and the Duke of York Islands. He regarded it as possibly conspecific with *S. modestus,* which occurs throughout much of northwestern New Guinea and the Admiralty Archipelago.

Tropidonophis hypomelas (2)—This snake was observed only twice. In both instances individuals were active during the day on the ground in open grassy vegetation along logging roads, less than 100 m from a river or stream. The species is endemic to Umboi, New Britain, Duke of York Islands, and New Ireland (Malnate and Underwood 1988).

Elapidae

Aspidomorphus muelleri (1)—Local villagers reported that this species was common in the area. The only individual of this venomous species that we saw was collected from an open forest/plantation area adjacent to the coast near Silur. This individual was aggressive and repeatedly attempted to bite. The species ranges throughout much of New Guinea and the Bismarck Archipelago (McDowell 1967).

Laticaudidae

Laticauda laticauda—Not collected during our survey. This widespread marine snake is known from India, Southeast Asia, including the Philippines, across to New Guinea and the Solomon Islands.

Typhlopidae

Acutotyphlops subocularis (1)—The single individual seen was collected from leaf litter in closed canopy forest following a heavy rain near the Weitin Base Camp. This blind snake is endemic to the Bismarck Archipelago and the Solomon Islands (McDowell 1974).

Ramphotyphlops depressus (2)—Both individuals were collected at night on the floor of closed canopy forest in the Weitin Valley lowlands after a heavy rain in the afternoon and early evening. *Ramphotyphlops depressus* is known from islands of eastern Papua New Guinea, including Manus, New Britain, New Ireland, St. Matthias, and the Duke of York Islands, and the Solomon Islands (McDowell 1974; Wallach 1996; McDiarmid et al. 1999).

Typhlops depressiceps (1)—The only individual that we observed was collected at night after heavy rain in closed canopy forest climbing on a rotten, mossy tree stump near the Weitin Base Camp. This is a new record for New Ireland. The species was previously known only from a few specimens collected from the north coast of New Guinea between the Sepik River and Lae (McDowell 1974), from the Port Moresby area (Menzies, pers. comm.), and more recently from New Britain (Foufopoulos 2001).

Discussion

The collecting methods that we used—hand collecting with assistance from local villagers and the use of rubber bands to stun small lizards—appear to have been very effective at inventorying the herpetofauna. In less than a month we worked out of three main field camps and obtained 39 of the 49 (79%) amphibian and reptile species now known from New Ireland. This included all but one of the eight frog species (88%), 20 of the 27 lizards (74%), and 12 of the 13 snakes (92%). It is possible that some species reported from New Ireland do not occur in the Weitin Valley and surrounding coastal areas, in which case our coverage of the herpetofauna of this region would be even higher.

Our survey brings the total number of amphibians and reptiles reported from New Ireland to 49. There are at least six additional wide-ranging reptile species that occur on surrounding islands and may occur on New Ireland (Gekkonidae: *Hemiphyllodactylus typus;* Scincidae*: Cryptoblepharus poecilopleurus, Emoia nigra, Eugongylus rufescens;* Typhlopidae*: Ramphotyphlops braminus;* Acrochordidae: *Acrochordus granulatus*. In addition there may be a few undescribed species that await discovery. However, it seems unlikely that the overall species total will grow to exceed 60–65 species.

All but three of the species that we collected had previously been reported from New Ireland. The three new records are *Platymantis browni* (new species described by Allison and Kraus 2001), *Sphenomorphus* sp., and *Typhlops depressiceps*. The taxonomy of *Sphenomorphus* is exceedingly complex; the single specimen of *Sphenomorphus* sp., the large species that we collected in the *jobiensis* complex, may represent a new taxon or a new locality record of an existing species (possibly *S. megaspilus* which is known from the Duke of York Islands). The discovery of *Typhlops depressiceps* in New Ireland was completely unexpected and illustrates the need for further study of the herpetofauna of the Papuan region.

Previous to our study, there were very few amphibian and reptile species reliably reported from southern New Ireland. The 39 species that we obtained from the Weitin Valley confirm that this region is an important center of diversity in

New Ireland. It is likely that the diversity of this region is rivaled only by the Lelet Plateau (one of the most visited upland collecting sites in New Ireland because it is easily accessible). However, we have not worked in the Lelet Plateau and do not feel that we can make a reliable comparison with that area on the basis of existing information.

The herpetofauna of New Ireland is very similar to that of New Britain. Only the ranid frog *Platymantis magnus,* the skink *Lipinia rouxi,* and possibly the large skink, *Sphenomorphus* sp. in the *jobiensis* complex, have not been reported from New Britain. However, many of the groups represented on New Britain seem not to have reached New Ireland. We looked intensely, without success, for *Tribolonotus.* However, all known species in this genus have rather secretive, semi-fossorial, habits and even if one occurred in the Weitin Valley, it is possible that we could have missed it. Mys (1988) searched without success for *Tribolonotus* in the Lelet Plateau and other areas in northern New Ireland; local people on the north coast reported to him that a species apparently occurred in their area. The microhylid frogs would be more difficult to miss but we recorded none. Most species, including those from New Britain, are easily located by their calls. However, neither did we find any *Rana kreffti,* which is reportedly found in New Ireland and should have been calling. Possibly these taxa are indeed absent from southern New Ireland.

The frog fauna of New Ireland seems depauperate. The island of Bougainville is about the same size as New Ireland (ca. 10,000 km^2) but has at least 25 species of frogs (Frost 1985). However, New Britain, four times larger than New Ireland, has only 13 species of frogs (Frost 1985). The New Britain fauna is poorly known, and it is likely that this number will increase substantially with additional collecting and study, particularly in montane areas. As discussed previously, we do not expect the frog fauna of New Ireland to increase much beyond the eight species currently known.

The lizard fauna also seems depauperate when compared to surrounding regions, the snakes less so. Bougainville has nearly 40 species of lizards, many of them endemic, compared to 26 species from New Ireland. However, New Britain has only ca. 30 species of lizards, although, again, this number will undoubtedly increase when the fauna is better known. There are ca. 15 species of snakes on New Britain, compared to 13 species on New Ireland and ca. nine species on Bougainville (McCoy 1980, Allison 1993, 1996).

The depauperate nature of New Ireland probably has a geological explanation. The area north of the Weitin River, together with New Hanover and the Admiralty Archipelago, today comprise a small plate fragment, the North Bismarck Plate. The southern tip of New Ireland, together with New Britain and what is today the Adelbert Mountains and the Huon Peninsula on the North Coast of New Guinea, comprise another plate fragment, the South Bismarck Plate (Taylor 1979; Pandolfi 1993). The platforms underlying these plates probably formed from elements of the Vitiaz Arc system during the Oligocene but did not become subaerial until late Miocene (10-8 MY) when there was considerable volcanism along the boundaries of the Australian and Pacific Plates (Kroenke 1984). At approximately 5-4 MY the South Bismarck Plate collided with New Guinea, forming the Huon Peninsula and Adelbert Mountains. At this time most of New Ireland, New Hanover, and the Admiralty Archipelago were approximately 500 km southeast of their present positions; they began moving northwest at about 3.5 MY in response to sea-floor spreading in the Manus Basin (Taylor 1979; Kroenke 1984). This means that New Ireland was much more isolated from New Britain in the past than it is today. In addition, most of New Ireland is covered with raised limestone (Dow 1977), evidence of past subsidence. Although evidence is sketchy, the original terrestrial biota of New Ireland may have disappeared when the area was under water in the late Pliocene.

Although we did not conduct a formal monitoring program, our simple assessments of the abundance of each species suggest that most have large populations in southern New Ireland. With respect to conservation, the species of greatest concern are generally those with the most restricted ranges. However, the two species of frogs endemic or nearly endemic to New Ireland, *Platymantis magnus* and *Platymantis browni*, were both common to abundant, as judged by the number heard calling during wet weather. Of the four additional amphibian and reptile species that are endemic to the Bismarck Archipelago, *Platymantis schmidti* was abundant, *Tropidonophis hypomelas* was seen only twice, *Emoia bismarckensis* was rare, and *Lipinia rouxi* was not recorded.

The abundance of snakes is notoriously difficult to assess, and *Tropidonophis hypomelas* may actually be more common than our two sightings would indicate. In any case, it occurs throughout the Bismarck Archipelago, and there is no reason to expect that populations in southern New Ireland face greater threats to survival than do those elsewhere.

Emoia bismarckensis does appear to be rare. We collected only two specimens, one from the Lake Camp (1180 m) and the other from the Riverside Camp (ca. 150 m). Both individuals came from leaf litter on the floor of closed canopy forests, the same habitat occupied by several other species of *Emoia*, so it is difficult to identify why *E. bismarckensis* is rare. It occurs elsewhere in New Ireland, as well as on New Britain, and from the scanty information available, seems to be rare throughout its range.

Lipinia rouxi, endemic to New Ireland, is known from only two localities: Fissola, a coastal settlement on the north east coast, and at ca. 950 m on the Lelet Plateau in the central part of the island. This suggests that the species has a wide elevational tolerance. However, we failed to find it in the Weitin Valley and it may not occur there.

We collected only a single specimen of a large species of *Sphenomorphus* in the *S. jobiensis* complex. This may represent an endemic taxon with a limited geographic range and deserves special attention.

Most of the remaining species, for which we have specimens, are both common and widespread. The exceptions are species for which we have 1 or 2 specimens and for which we could not obtain reliable information on abundance from the local people. Most of these represent fossorial or cryptic species with extensive ranges in the Papuan region. These include two skinks, *Sphenomorphus wolfi, Eugongylus albofasciolatus,* and the blind snakes, *Ramphotyphlops depressus, R. subocularis,* and *Typhlops depressiceps.* These species are generally difficult to find and we were unable to assess their abundance. *Typhlops depressiceps* deserves special attention because it was previously thought to be restricted to mainland New Guinea. The other species are common elsewhere.

The Bismarck Boa, *Bothrochilus boa,* is an extremely valuable species in the pet trade, and reptile smugglers have been known to collect large numbers for illicit sale on the black market (someone apprehended ca. two years ago by Australian authorities had more than 40 specimens). Although this snake was reasonably common in the Weitin Valley, local populations could easily disappear due to overcollecting. Wildlife authorities need to be especially vigilant and to work closely with the local people to deal with this problem.

Acknowledgments

We would like to thank the other members of the expedition, especially Phil Angle, Sam Antiko, Bruce Beehler, Louise Emmons, Robin Foster, Matthew Goga, Michael Hedemark, Felix Kinbag, Tommy Kosi, Michael Lucas, Robert and Karen McCallum, Larry Orsak, Wayne Takeuchi, Joseph Wiakabu, as well as our team of field collectors, especially Greg Langman, Isador Tosemi, Stanis Todiring, and Koniel Tuarianu. We are also especially grateful to Bruce Beehler, Mike Hedemark, John Geno, and Bruce Jefferies for logistical arrangements. Carla Kishinami and Terry Lopez deserve special thanks for the skill and efficiency with which they processed our collection at Bishop Museum. We also thank Leeanne Alonso, Bruce Beehler, Aaron Bauer, Robert Cowie, Lu Eldredge, and Carla Kishinami for reviewing earlier drafts of this report and Addison Winn for identifying the blind snakes. To the people of the Weitin Valley, we tender our profound thanks for their help, assistance and friendship during the expedition, for permission to work on their land, and for their interest in the conservation of biological diversity.

Literature Cited

Allison, A. 1993. Biodiversity and conservation of the fishes, amphibians, and reptiles of Papua New Guinea. *In* B. M. Beehler (ed.), Papua New Guinea conservation needs assessment. Volume 2. Vol. 2, p. 157–225. The Biodiversity Support Program, Washington, DC.

Allison, A. 1996. Zoogeography of amphibians and reptiles of New Guinea and the Pacific region. *In* A. Keast and S.E. Miller (eds.), The origin and evolution of Pacific Island biotas, New Guinea to eastern Polynesia. pp. 407–436. SPB Academic Publishing, Amsterdam.

Allison, A. and F. Kraus. 2001. New species of frog of *Platymantis* (Anura: Ranidae) from New Ireland. *Copeia* 2001(1):194–202.

Bauer, A.M. 1994. Familia Gekkonidae (Reptilia, Sauria). Part 1 Australia and Oceania. Das Tierreich 109:xiii + 1–306.

Bauer, A. M. and G.J. Watkins-Colwell. 2001. On the origin of the types of Hypsilurus godeffroyi (Reptilia: Squamata: Agamidae) and early German contributions to the herpetology of Palau. *Micronesica* 34(1):73–84.

Bailey, P. 1976. Food of the marine toad, *Bufo marinus*, and six species of skink in a cacao plantation in New Britain, Papua New Guinea. Australian Wildlife Research 1976(3):185–188.

Beehler, B.M. (ed.) 1993. A Biodiversity Analysis for Papua New Guinea. Conservation Needs Assessment. Volume 2. Biodiversity Support Program. Washington, DC, USA.

Brown, W.C. 1983. A new species of *Emoia*, Sauria (Scincidae), from New Britain. *Steenstrupia* 8(17):317–324.

Brown, W.C. 1991. Lizards of the genus *Emoia* (Scincidae) with observations on their ecology and biogeography. Memoirs of the California Academy of Sciences 15:1–94.

Brown, W.C., and J.I. Menzies. 1978 (1979). A new *Platymantis* (Amphibia: Ranidae) from New Ireland with notes on the amphibians of the Bismarck Archipelago. Proceedings of the Biological Society of Washington 91(4):965–971.

Brown, W.C., and M.J. Tyler. 1968. Frogs of the genus *Platymantis* (Ranidae) from New Britain with descriptions of new species. Proceedings of the Biological Society of Washington 81:69–86.

Brown, W.C., and T.P. Webster. 1969. A new frog of the genus *Discodeles* from Guadacanal Island. *Breviora* 338:1–9.

Bruna, E.M., R.N. Fisher and T.J. Case. 1995. Cryptic species of Pacific skinks (*Emoia*): further support from mitochondrial DNA sequences. *Copeia* 1995(4): 981: 983.

Cogger, H.G. 1972. A new scincid lizard of the genus *Tribolonotus* from Manus Island, New Guinea. *Zoologische Mededelingen (Leiden)* 47:201–210.

Cogger, H.G. 1992. Reptiles and amphibians of Australia. Fifth ed. Cornell University Press, Ithaca. 765 p.

Donnellan, S.C. and K.P. Aplin. 1989. Resolution of cryptic species in the New Guinean lizard *Sphenomorphus jobiensis* (Scincidae) by electrophoresis. *Copeia* 1989(1): 81–88.

Dow, D.B. 1977. A geological synthesis of Papua New Guinea. Bureau of Mineral Resources, Geology and Geophysics Bulletin, Australian Government Publishing Service, Canberra 201:vii + 41.

Foufopoulos, J. 2001. *Typhlops depressiceps. Herpetological Review* 32(1):61–62.

Frost, D. 1985. Amphibian species of the world, a taxonomic and geographic reference. Allen Press and Association of Systematics Collections, Lawrence, Kansas. 732 p.

Greer, A.E. 1970. The relationships of the skinks referred to the genus *Dasia*. Breviora 348:1–30.

Greer, A.E., and F. Parker. 1974. The *fasciatus* species group of *Sphenomorphus* (Lacertilia: Scincidae): notes on eight previously described species and descriptions of three new species. Papua New Guinea Scientific Society Proceedings 25:31–61.

Greer, A.E., and B. Mys. 1987. Resurrection of *Lipinia rouxi* (Hediger, 1934) (Reptilia: Lacertilia: Scincidae), another skink to have lost the left oviduct. Amphibia-Reptilia 8:417–426.

Günther, A. 1877. On a collection of reptiles and fishes from Duke of York Island, New Ireland, and New Britain. Proceedings of the Zoological Society of London 1877:127–132.

Hamilton, W. 1979. Tectonics of the Indonesian region. U.S. Geological Survey Professional Paper 1078: 1–345.

Hediger, H. 1934. Beitrag zur Herpetologie und Zoogeographie Neu-Britanniens und einiger umliegender Gebiete. Zoologische Jahrbücher Abtteilung für Systematik 65(5/6):441–582.

Ineich, I., and G.R. Zug. 1991. Nomenclatural status of *Emoia cyanura* (Lacertilia, Scincidae) populations in the central Pacific. *Copeia* 1991(4):1132–1136.

King, F.W. and R.L. Burke. 1989. Crocodilian, tuatara and turtle species of the world. Association of Systematics Collections, Washington, DC. 216 p.

Kluge, A.G. 1993. *Aspidites* and the phylogeny of pythonine snakes. Records of the Australian Museum Supplement 19:1–77.

Kroenke, Loren W. 1984. Cenozoic tectonic development of the Southwest Pacific. United Nations Economic and Social Commission, Committee for Co-Ordination of Joint Prospecting for Mineral Resources in South Pacific Offshore Areas (CCOP/SOPAC), Technical Bulletin 6:1–122.

Malnate, E.V., and G. Underwood. 1988. Australasian natricine snakes of the genus *Tropidonophis*. Proceedings of the Academy of Natural Sciences of Philadelphia 140(1):59–201.

McCoy, M. 1980. Reptiles of the Solomon Islands. Wau Ecology Institute. Handbook 7, Wau, Papua New Guinea. 82 p.

McDiarmid, R.W., J.A. Campbell and T.A. Touré. 1999. Snake species of the world: a taxonomic and geographic reference. Vol. 1. Herpetologist's League, Washington, DC. 511 p.

McDowell, S.B. 1967. *Aspidomorphus*, a genus of New Guinea snake of the Family Elapidae, with notes on related genera. Journal of Zoology, London, 151:497–543.

McDowell, S.B. 1972. The species of *Stegonotus* (Serpentes, Colubridae) in Papua New Guinea. Zoologische Mededelingen (Leiden) 47:6–26.

McDowell, S.B. 1974. A catalogue of the snakes of New Guinea and the Solomons, with special reference to those in the Bernice P. Bishop Museum. Part I. Scolecophidia. Journal of Herpetology 8(1):1–57.

McDowell, S.B. 1975. A catalogue of the snakes of New Guinea and the Solomons, with special reference to those in the Bernice P. Bishop Museum. Part II. Aniliodea and Pythoninae. Journal of Herpetology 9(1):1–79.

McDowell, S.B. 1979. A catalogue of the snakes of New Guinea and the Solomons, with special reference to those in the Bernice P. Bishop Museum. Part III. Boinae and Acrochordoidea. Journal of Herpetology 13(1):1–92.

Menzies, J.I. 1982a. Systematics of *Platymantis papuensis* (Amphibia: Ranidae) and related species in the New Guinea region. British Journal of Herpetology 6:236–240.

Menzies, J.I. 1982b. The voices of some male *Platymantis* species of the New Guinea region. British Journal of Herpetology 6:241–245.

Menzies, J.I. 1987. A taxonomic revision of the Papuan *Rana* (Amphibia: Ranidae). Australian Journal of Zoology 35:373–418.

Mys, B. 1988. The zoogeography of the scincid lizards from the north Papua New Guinea (Reptilia: Scincidae). I. The distribution of the species. Bulletin Institut Royal des Sciences Naturelles Belgique Biologie 58:127–184.

O'shea, Mark. 1996. A guide to the snakes of Papua New Guinea. Independent Publishing, Port Moresby, Papua New Guinea. 239 p.

Pandolfi, J.M. 1993. A review of the tectonic history of New Guinea and its significance for marine biogeography. Proceedings of the Seventh International Coral Reef Symposium, Guam 2:718–728.

Philipp, Kai M., Wolfgang Böhme and Thomas Ziegler. 1999. The identity of *Varanus indicus:* redefinition and description of sibling species coexisting at the type locality. Spixiana 22(3):273–287.

Ramsay, E.P. 1878. Description of a new species of *Pelodryas* from New Ireland. Proceedings of the Linnean Society of New South Wales 2(1):28–30.

de Rooij, N. 1917. The reptiles of the Indo-Australian Archipelago. II. Ophidia. E.J. Brill, Leiden. 334 pp.

Taylor, B. 1979. Bismarck Sea: evolution of a back-arc basin. Geology 7(4):171–174.

Tyler, M. J. 1964. Transfer of the New Britain frog *Hyla brachypus* (Werner) to the microhylid genus *Oreophryne*. Mitteilungen aus dem Zoologischen Museum im Berlin 40(1):3–8.

Tyler, M.J. 1967. Microhylid frogs of New Britain. Transactions of the Royal Society of South Australia 91:187–190.

Tyler, M.J. 1968. Papuan hylid frogs of the genus *Hyla*. Zoologische Verhandelingen 96:3–203.

Waite, E.R. 1897. A new blind snake from the Duke of York Island. Records of the Australian Museum 3(3):69–70.

Wallach, V. 1996. The systematic status of the *Rampho-typhlops flaviventer* (Peters) complex (Serpentes: Typhlopidae). *Amphibia-Reptilia* 17(4):341–359.

Wermuth, H. 1967. Lizte der rezenten Amphibien und Reptilien: Agamidae. Das Tierreich 86:1–127.

Werner, F. 1898. Vorläufige Mittheilung über die von Herrn Prof. F. Dahl im Bismarck-archipel gesammelten Reptilien und Batrachier. Zoologischer Anzeiger 21:552–556.

Werner, F. 1900. Die Reptilien und Batrachierfauna des Bismarck-Archipels. Mitteilungen aus dem Zoologischen Museum im Berlin 1:1–132.

Zug, G.R., E. Lindgren, and J.R. Pippet. 1975. Distribution and ecology of the marine toad, *Bufo marinus*, in Papua New Guinea. Pacific Science 29(1):31–50.

Zweifel, R.G. 1960. Results of the 1958–1959 Gilliard New Britain Expedition. 3. Notes on the frogs of New Britain. American Museum Novitates 2023:1–27.

Zweifel, R.G. 1966. A new lizard of the genus *Tribolonotus* (Scincidae) from New Britain. American Museum Novitates 2264:1–12.

Zweifel, R.G. 1975. Two new frogs of the genus *Platymantis* (Ranidae) from New Britain. American Museum Novitates 2582:1–7.

Zweifel, R.G. 1979. Variation in the scincid lizard *Lipinia noctua* and notes on other *Lipinia* from the New Guinea region. American Museum Novitates 2676:1–21.

Chapter 5

A Field Survey of the Resident Birds of Southern New Ireland

Bruce M. Beehler, J. Phillip Angle, David Gibbs, Michael Hedemark, and Daink Kuro

Abstract

Combining the records of field trips in 1993 and 1996 to southern New Ireland, we recorded 89 species of breeding land and freshwater bird species. The New Britain Sparrowhawk (*Accipiter brachyurus*) is recorded on New Ireland for the first time. Our sight record of a probable Meyer's Goshawk (*Accipiter meyerianus*), if confirmed, will constitute a second forest raptor new to New Ireland's list, both species being known to inhabit neighboring New Britain. Finally, David Gibbs recorded a probable Marbled Frogmouth (*Podargus ocellatus*) vocalizing, also which would be new to the island.

In spite of intensive searches, we were unable to confirm the presence in the current avifauna of a large forest rail, a *Zoothera* ground-thrush, a *Cichlornis* thicket-warbler, and a medium-sized honeyeater, four species expected on distributional grounds. New Ireland's avifauna is characterized by its impoverishment, possibly a product of both human activities (see Steadman 1995) and other past environmental events that also appear to have influenced the herpetofauna.

Introduction

It is safe to say that the avifauna of New Ireland has been relatively neglected by ornithologists over the last century, especially when compared to the number of studies that have been conducted on the nearby mainland of New Guinea (see Mayr and Diamond 2001, the *Literature Cited* sections of Coates 1985, 1990). Part of the reason for this neglect is that New Ireland is a true oceanic island (Diamond 1970) with an avifauna that is considerably poorer than that of New Guinea. Nonetheless, because of the low human population, the extensive forest cover, the presence of forested mountain ranges exceeding 2000 meters elevation, and the general lack of recent field surveys of the avifauna using the most up-to-date techniques, there was reason to expect at the time of our expedition the avifauna of New Ireland had not been completely described.

In addition, the presence on neighboring Bougainville and New Britain islands of local representatives of a large forest rail, a *Zoothera* ground-thrush, a *Cichlornis* thicket-warbler, and a medium-sized montane honeyeater—none of which had yet been recorded from New Ireland—offered support for this notion. Steadman's (1995, and pers. comm.) ongoing delineation of New Ireland's prehistoric avifauna, based on his study of bird bones from archaeological sites that include at least eight species of birds either new to science or not previously recorded from New Ireland, presented us with an anomaly that we wished to address. Were some of Steadman's missing species still lurking, undetected, in some unsurveyed habitat on New Ireland?

One final question begged an answer. To what extent were the large-scale logging operations being conducted in southern New Ireland posing a threat to the future of the birds?

Previous Research

Most early studies of the birds of New Ireland and the neighboring Duke of York Islands and New Britain were conducted by early pioneering voyagers (Lesson 1828), by German colonial naturalists (e.g., Dahl 1899, Reichenow 1899, Heinroth 1902, 1903), or through the initiative of English naturalists (Sclater 1877, Ramsay 1878, Hartert 1925) or German missionaries (Meyer 1936). These early studies focused almost exclusively on collections of specimens assembled by professional field collectors working on behalf of European or Australian institutions or wealthy patrons of natural history. Collections of study skins arrived in museums, and from these troves taxonomists diagnosed and described new species and races—a major and important goal of the field at that time. These efforts brought to light all five of New Ireland's known endemic bird species: the White-naped Lory (*Lorius albidinuchus*), the Long-tailed Drongo (*Dicrurus megarhynchus*), the New Ireland Mannikin (*Lonchura forbesi*), the New Ireland Myzomela (*Myzomela pulchella*), and the New Ireland Friarbird (*Philemon eichhorni*). More recently, additional attention has been given to New Britan (Anthony et al., not dated; Bishop and Jones 2001).

Most recently, Salomonsen (1962, 1964, 1965), Beehler (1978), Finch and McKean (1987), and Jones and Lambley (1987) have paid brief visits to New Ireland, either to the Lelet Plateau, selected fringing islands, or the southeastern lowlands and adjacent interior highlands. Salomonsen described a new monarch *Monarcha ateralba* from nearby Djaul Island; Beehler added the widespread White-throated Pigeon (*Columba vitiensis*) and Island Thrush (*Turdus poliocephalus*) to the New Ireland avifauna; and Finch and McKean noted an unidentified goshawk and an unidentified flycatcher.

Methods

Our report combines the results of the RAP survey with those of the third author, David Gibbs (DG), who independently surveyed birds in southern New Ireland in 1993. Gibbs surveyed birds along a transect from Silur and Taron on the coast up through the hills to the ridge that linked the Lake and Top camps of the 1994 RAP survey. This is essentially the exact route of the survey made by Beehler (1978). Gibbs' highest observations were made at 1700 m.

From 15 January–15 February, the RAP team surveyed birds at the four main camps (see Gazetteer): Weitin Base Camp (240 m) 9 days, Lake Camp (1200 m) 11 days—January 24–31; Top Camp (1800 m) 6 days, and Riverside Camp (160 m) 12 days. Additional ad lib observations were made at a few coastal localities as time and opportunity permitted.

We surveyed birds using a range of methods, some of which were appropriate to a general faunal survey, others of which were focused on estimating relative abundance of species in particular habitats. First and most generally, we relied on sight and sound encounters to locate and identify species whenever we were afield. Records were compiled daily of birds observed, netted, and collected at each camp.

Mist-nets were used to sample forest birds. The nets (12 meter length, 36 and 61 mm mesh sizes) were set primarily at ground level, but in a few instances were pulleyed into the middle story of the forest interior to capture species that rarely flew in the lower levels of the forest. All trapped birds were identified, measured, and recorded on data sheets. Because of the rarity of museum material from New Ireland, a sampling of these birds was preserved as museum study skins, skeletal specimens, or fluid specimens. Prepared almost wholly by JPA, these are now stored in the National Museum of Natural History (Washington, DC, USA) and the PNG National Museum and Art Gallery (Waigani). Others were measured and then released after the tips of two rectrices were clipped as a visible marker.

Various members of the expedition party took turns carrying a shotgun into the field to collect voucher specimens of hard-to-net birds and mammals. This was used to provide unambiguous records of either little-known or hard-to-identify species.

Sight-and-sound censuses were carried out daily by BMB at each camp. Each 45-minute census was initiated either at 0630 or 1530 hrs (weather permitting). A fixed circuit route was followed, and all birds detected within 150 m of the censuser were recorded. The recorder followed a pre-determined circuit route, moving slowly and steadily unless it was necessary to stop to either locate or identify birds. Series of censuses were completed at all except for the Weitin Base Camp. For analysis, we compare the combined results from seven censuses (five morning and two afternoon censuses) from each site (this is the number that were completed at the Top Camp).

Results

We recorded 89 species of land and freshwater breeding birds (Appendix 4), including three species new to New Ireland, the goshawks *Accipiter meyerianus* and *Accipiter brachyurus* and the Marbled Frogmouth *Podargus ocellatus*. The goshawks are known from New Britain, and the frogmouth is expected to occur on New Britain on distributional grounds (it is recorded from Bougainville Island and mainland New Guinea). We failed to find evidence for New Ireland populations of a large forest rail, *a Zoothera* ground-thrush, a *Cichlornis* thicket-warbler, or a medium-sized honeyeater (examples of these inhabit New Britain and Bougainville). In addition, we did not encounter any flycatchers that might correspond to the undescribed *Microeca* that was observed by Finch and McKean (1987).

With regard to the RAP surveys, the audial censuses located 58 species, netting produced 208 individuals of 31 species, and hunting produced 18 individuals of 14 species. One hundred and seventy-three (173) voucher specimens of 38 species were prepared for museum collections. These comprised 50 skeletal specimens, 85 fluid specimens, and 38 study skins.

Both netting and census results show that the forest avifaunas are impoverished both in species numbers and numbers of individuals (Appendix 5). In this manner, New Ireland is typical of an oceanic Pacific island (Diamond

1970). In spite of New Ireland's considerable topographical and ecological heterogeneity, the forest birdlife is poorly represented when compared to that on mainland New Guinea (Diamond 1970, Beehler 1981, Beehler et al. 1995).

Census results (Appendix 5) show that species richness declines with elevation (lowlands 42 spp/346 individuals; 1200 m: 20 spp/180 individuals; 1800 m: 16 spp/216 individuals). The two upland census routes produced considerably fewer individuals than the single lowland route. In both upland census datasets (Appendix 6) virtually the entire "middle section" of the avifauna (cuckoos, swifts, and Coraciiformes) was absent. For some reasons yet to be determined, these groups are primarily lowland-dwellers on New Ireland.

We recorded all five of New Ireland's known endemic species. *Lonchura forbesi* is a grassland estrildid mannikin that we recorded only at the Silur airstrip clearing. It is presumably present through the coastal strip where grassland openings are common. The four forest-dwelling endemics include one elevationally widespread form (the drongo *Dicrurus megarhynchus*) and three montane forms (the lory *Lorius albidinuchus*, the friarbird *Philemon eichhorni*, and the small honeyeater *Myzomela pulchella*). The drongo was common on both the Riverside and Lake camp censuses (Riverside: 13 indiv., 6 censuses; Lake: 8 indiv., 5 censuses). The myzomela honeyeater was abundant at the Top Camp (24 indiv., 7 censuses). The friarbird was abundant at the Top Camp (16 indiv., 7 censuses) and at the Lake Camp (21 indiv., 7 censuses). The lory was recorded solely on the Lake Camp census (2 indiv., 1 census).

Selected species accounts

Little Grebe

Tachybaptus ruficollis—A pair in breeding plumage was observed daily on the pond at the Lake Camp, apparently resident there. Vocal, active, and quite inquisitive.

White-bellied Sea-Eagle

Haliaeetus leucogaster—Found foraging in the uplands of the Hans Meyer Range. A juvenile was encountered feeding on the ground on a cuscus at 1700 m in mossy forest interior (positively identified by a shed feather that was confirmed by Roxie Laybourne of the NMNH, Washington, DC).

Meyer's Goshawk

Accipiter meyerianus—A single sighting of a very large goshawk, ascribed to this species was made by MH in the forest at Top Camp.

Bismarck [New Britan] Sparrowhawk

Accipiter brachyurus—Encountered at the Lake and Top camps. Apparently widespread. It is highly likely that this was the species that Finch and McKean (1987) observed and misidentified as *A. luteoschistaceus*.

Yellow-legged Pigeon

Columba pallidiceps—Not observed by us in 1993 or 1994, by Beehler (1978), or by Jones and Lambley (1987). Finch and McKean (1987: 6) report "Only one bird seen, over forest near Taron, where the hills come almost to the sea. This is a strikingly rare species in southern New Ireland and merits attention. It is virtually unknown throughout its range (Goodwin 1983, Coates 1985), although it is apparently not rare on Makira, Solomon Islands (Roger James verbal communication to Guy Dutson 1998).

Finsch's Imperial Pigeon

Ducula finschii—Observed by DG in 1993 at 750 m, but not recorded by the 1994 effort, nor by Finch and McKean or Jones and Lambley. Beehler (1978) observed a single individual. Apparently common if the vocalization is recognized (G. Dutson, in litt.).

White-naped Lory

Lorius albidinuchus—Fairly common singly or in pairs in the middle elevations. Recorded on seven of seven days at the Lake Camp (1200 m)—but absent from all but one formal census. Not recorded at the Top Camp, nor from the lowland camps.

Buff-faced Pygmy-Parrot

Micropsitta pusio—J.M. Diamond (in litt.) confirms that this species is incorrectly included on the New Ireland list. There is no place where two species of pygmy-parrots share the forest in true sympatry.

Torresian/Bismarcks Crow

Corvus [orru/insularis]—The New Ireland *Corvus* has traditionally been treated as *C. orru* (Mayr 1955, Coates 1990), but we agree with Finch and McKean (1987) that the crow on New Ireland and New Britain is vocally very distinct from that of the mainland New Guinea *C. orru*. Given the great importance of vocalizations in this morphologically invariant genus, we suggest that the Bismarcks populations may represent a sibling species distinct from the populations of New Guinea and Australia. This issue merits additional attention.

Discussion

Lost Birds

Ornithologically, New Ireland is remarkable more for what is absent from the modern avifauna than for what is present. The five endemic species are significant, but this count is low when compared to that for neighboring New Britain (16 endemic species). Most remarkable are New Ireland's five "missing" species—lineages for which island representatives occur both on New Britain and Bougainville: a rail, a white cockatoo, a ground-thrush, a thicket-warbler, and a medium-sized honeyeater. These can be likened to missing foundation-stones in a building. They should be present, and without them there is evidence that the avifauna may be significantly impoverished, at least in the context of local island patterns of community composition.

Where did these birds go? Or, were they ever on New Ireland in the first place? Biogeographically it is hard to imagine these species colonizing Bougainville and New Britain without also colonizing intervening New Ireland. David Steadman's work (1995) with the bird bones recovered from the Balof archeological sites in central New Ireland provide a preliminary answer—one that at this time must suffice. His research has turned up two of the "lost" birds—a cockatoo and a rail (both large species that might be expected to be prominent in a bone midden). Thus we now have clear evidence that two of our "lost" five *did* inhabit New Ireland. Steadman has not yet studied the prehistoric passerine bones from New Ireland, so in fact evidence for the former presence of the ground-thrush, thicket-warbler, and medium-sized honeyeater may yet be forthcoming. We believe that all five of the missing species do or did, at least formerly, inhabit New Ireland.

We have no doubt that the cockatoo is gone. The rail, we believe, might possibly be lurking in a swamp or forest, awaiting detection by a future researcher – Guy Dutson received reports of a flightless rail in the Weitin Valley. The thrush, warbler, and honeyeater are all typical of species that are often caught in mist-nets, and the three tend to be vocal songsters. We doubt they remain on New Ireland, but challenge future ornithologists to prove us wrong. The area surveyed in this effort was a tiny sliver of the habitat available for rare or localized species.

The mystery *Microeca* flycatcher that Finch and McKean sighted on New Ireland and New Britain poses a real conundrum. Coates (1990) reports that the bird was also observed by T. Palliser in 1987 near to where it was seen by the original group—in montane forest not far from our Lake Camp. In addition, Guy Dutson has observed this bird. We saw nothing like it in our surveys so can provide little to solve this mystery. The most wide-ranging species of *Microeca* appears to be *M. flavigaster*. We suggest either that there may be a migratory population of *M. flavigaster*, or else that the species is expanding its range eastward from New Guinea. Its patchy range in northern New Guinea implies an expanding population, moving northward from its original Australian distribution.

An Empty Avifauna

An ornithologist comparing the birdlife of a mossy montane forest above the Tari valley of mainland New Guinea with that of the forest at our Top Camp on New Ireland would be shocked by the contrast. Although the New Ireland forest would look a lot like the mainland forest, it would not sound like the mainland forest. It would be nearly silent, except for the occasional notes of a friarbird, myzomela, leaf warbler, or pigeon. Dawn chorus is practically absent from the upland forests of New Ireland. This is because the upland forest avifauna is so depauperate. The hill and lowland forests are similarly impoverished by comparison to their mainland counterparts. There are fewer species and the individual species have not increased in numbers to fill the ecological "space" left by the missing species.

As indicated by the data from the other chapters, there is less of everything, and what appears to be a similar forest environment is, in fact, attenuated taxonomically in all sorts of ways—fewer species of trees, arthropods, mammals, etc. One potential explanation is impacts of massive volcanism from the various active sites surrounding New Ireland. Huge ashfalls could, perhaps, have devastated the forests and the forest avifauna, with some particularly poor dispersers failing to recover. It would be worthwhile to examine the archeological or soil records for evidence of major eruptions.

Threats to the Avifauna

Southern New Ireland remains largely forested. The current logging by foreign national operations is extracting selectively, but intensively and with high impact trees, from the lowlands and lower hills. No other large-scale development is planned (except the influences of the Lihir mine might reach this area, a least indirectly). To what degree is the avifauna threatened by these human activities? This is not easy to answer, in part because of the information regarding "lost" species. Steadman's archaeological evidence supports the notion that New Ireland's prehistoric human inhabitants may have promoted the extirpation of entire island populations of eight or more species (a large heron, an ibis, a rail, a giant swamphen, a cockatoo, a tytonid owl, a typical owl, and a large crow). Such seems plausible on a tiny oceanic island, but how can this sort of extirpation occur on an island as large and rugged as New Ireland? We can only guess that, as happened on the Hawaiian islands, the local human population focused on harvesting these species for some reason (for food, for feathers, or for other traditional uses) or that these species were especially vulnerable to slightly less direct forms of prehistoric human impact, such as partial deforestation, predation/competition from introduced species of birds and mammals, or non-native pathogens. The cause of these extinctions remains a mystery of considerable significance, mainly because it demonstrates graphically how vulnerable forest birds can be to human influences. One would never have predicted the extirpation of a cockatoo from an island as large, mountainous, and forested as New Ireland in advance of western colonial contact. This fact must make us wary of stating that any particular bird species is "safe." We simply do not know all the kinds of vulnerability that exist. Some of these species, for example, may have been eradicated by viruses or other pathogens imported along with the commensal fauna brought by the human settlers (a cuscus, a wallaby, a toad, among others).

Allison and Bigilale's (Chapter 4, this volume) discussion of "missing" skink species—taxa that are expected on New Ireland on biogeographic grounds—conforms to our findings on all but one important factor—one would be loath to argue that the skinks were "hunted out" by pre-contact human populations. If not from that, why are they absent? We would argue there may be another, non-anthropogenic answer that still awaits discovery. Some major natural environmental event might have struck this island at some point in the recent past. The discovery of such an event should be a goal for paleoecologists or paleobiologists.

Future Focus of Ornithological Research

As exemplified by the final point above, there are many productive topics for future field study in New Ireland:

1. Search for the "lost" species. We recommend thorough and focused searches in those areas yet unvisited, and applying, more intensively, the use of tape recorders and tape playbacks where appropriate. If recordings of related species can be assembled, the rail, the thrush, and the thicket-warbler are amenable to this sort of search.

2. Search for the "Microeca" flycatcher. It is important that voucher specimens be secured of this species, as it is clearly difficult to diagnose because of its size and nondescript plumage.

3. Work the wetland/swamp forest habitats. This is a likely hiding place for the missing rail (or even Steadman's giant swamphen).

4. A field reconnaissance is needed to the upland and summit forests of the Verron Range, south of the Weitin gap. Foster has shown (in this publication) that the vegetation in the Verron Range is distinct from that in the Hans Meyer Range, where all upland work has been focused. Might this distinct and isolated environment support a distinct bird fauna?

5. Visit the rugged upland plateau area of the Hans Meyer Range, just NNW of the highest summit, identified by the presence of a small deep-water crater lake. This isolated area seems to support a large expanse of undulating upland forest (especially the valley just SE of the lake) and could prove to be a stronghold of one or more of the "lost" species.

6. Monitor bird populations in pre-logged, logged, and post-logged forests, looking to document significant effects on bird populations. It would provide important data on the relative vulnerability of surviving species to habitat modification. Take particular notice of nesting trees for Blyth's Hornbill. We suspect the selective culling of large-boled trees will pose a threat to reproduction by this hornbill, which requires large trees with cavities for nesting.

7. Continue paleoecological research. Steadman's sample of prehistoric bird bones is much too small to be comprehensive. Additional coves and rock shelters should be excavated to obtain better data on missing species. Paleobotanical research (especially pollen cores from wetlands) would provide a history of human land-use patterns.

Acknowledgments

We thank David Steadman for providing information about his identification of bird bones from the Balof archaeological sites. Jared Diamond kindly provided distributional and bibliographic data on New Ireland birds, based on his monographic treatment of the region's avifauna, under preparation. We thank Mary LeCroy for comparing our specimen of *Accipiter brachyurus* with material in the American Museum of Natural History. Guy Dutson critically read the text and provided a number of useful comments and improvements. Roxie Laybourne kindly identified the eagle feather.

Literature Cited

Anthony, N., D. Byrnes, J. Foufopoulos, and M. Putnam, (not dated.). New Britan Biological Survey. MMI, University of Wisconsin, Madison, Wisconsin.

Beehler, B.M. 1978. Notes on the mountain birds of New Ireland. Emu 78: 65–72.

Beehler, B.M. 1981. Ecological structuring of forest bird communities in New Guinea. Monogr. Biol. 42: 837–862.

Beehler, B.M., J.B. Sengo, C. Filardi, and K. Merg. 1995. Documenting the lowland rainforest avifauna in Papua New Guinea—Effects of patchy distributions, survey effort and methodology. Emu. 95: 149–161.

Bishop, K.D., and D.N. Jones. 2001. The montane avifauna of West New Britan, with special reference to the Nakanai Mountains. Emu 101: 205–220.

Coates, B. 1985. Birds of Papua New Guinea. Vol 1. Dove Publications, Alderley, Queensland.

Coates, B. 1990. Birds of Papua New Guinea. Vol 2. Dove Publications, Alderley, Queensland.

Dahl, F. 1899. Das Leben der Vögel auf den Bismarckinseln. Mitt. Zool. Mus. Berlin 1: 107–222.

Diamond, J.M. 1970. Ecological consequences of island colonization by southwest Pacific birds, II. The effect of species diversity on total population density. Proc. Natl. Acad. Sci. USA. 67: 1715–1721.

Finch, B.W., and J.L. McKean. 1987. Some notes on the birds of the Bismarcks. Muruk 2: 3–28.

Goodwin, D. 1983. Pigeons and Doves of the World. Comstock Publishing Associates. Ithaca, New York.

Hartert, E. 1925. A collection of birds from New Ireland (Neu Mecklenburg). Novit. Zool. 32: 115–136.

Heinroth, O. 1902. Ornithologische Ergebnisse der 'I. Deutschen Südsee Expedition von Br. Mencke.' J. Ornith., Leipz. 50: 390–457.

Heinroth, O. 1903. Ornithologische Eergebnissse der 'I. Deutschen Südsee Expedition von Br. Mencke.' J. Ornith., Leipz. 51: 65–126.

Jones, D., and P. Lambley. 1987. Notes on the birds of New Ireland. Muruk 2: 29–33.

Lesson, R.P. 1828. Voyage Coquille. Zool. i. 1. In Voyage atour du Monde. L.I. Duperrey (ed.). Paris: A. Betrand.

Mayr, E. 1955. Notes on the birds of Northern Melanesia. 3—Passeres. American Museum Novitates 1707.

Mayr, E., and J. M. Diamond. 2001. The birds of Northern Melanesia. Oxford University Press, New York.

Meyer, O. 1936. Die Vogel des Bismarckarchipel. Vunapope P.O. Kokopo. Katholosche Mission, 55 pp.

Ramsay, E.P. 1878. Description of some new species of birds from New Britain, New Ireland, Duke of York Island, and the southeast coast of New Guinea. Proc. Linn. Soc. NSW 2: 104–107.

Reichenow, A. 1899. Die Vögel des Bismarckinseln. Mitteilungen des Zoologischen Museums Berlin 1(3): 1–106.

Salomonsen, F. 1962. Whitehead's Swiftlet (Collocalia whiteheadi Ogilvie-Grant) in New Guinea and Melanesia. Noona Dan Papers (3) Vidensk. Meddr. Dasnk. Naturh. Foren. 128: 78–83.

Salomonsen, F. 1964. Some remarkable new birds from Dyaul Island, Bismarck Archipelago, with zoogeographical notes. Biol. Skr. K. Dan. Vidensk. Selsk. 14: 1–37.

Salomonsen, F. 1965. Notes on the Mountain Leaf Warbler (Phylloscopus trivirgatus Strickland) in the Bismarck Archipelago. Vidensk. Meddr. Dansk. Naturh. Foren. 128: 78–83.

Sclater, P.L. 1877. On birds collected by Mr. George Brown, C.M.Z.S., on Duke of York Island, and on the adjoining parts of New Ireland and New Britain. Proc. Zool. Soc. London: 96–114.

Steadman, D.W. 1995. Prehistoric extinctions of Pacific island birds: biodiversity meets zooarcheology. Science 267: 1123–1131.

Chapter 6

Survey of Mammals of Southern New Ireland

Louise H. Emmons and Felix Kinbag

Abstract

In January–February 1994 we surveyed mammals in southern New Ireland along an elevational transect from the interior lowlands of the Weitin Valley to a summit ridge of the Hans Meyer Range at 1830 m. We recorded 26 species, including 17 bats, 3 marsupials, 3 rodents, and 3 feral domestic mammals. Four bat species are new records for the island: *Pteropus gilliardorum, Mosia nigrescens papuana, Nyctophilus microtis,* and *Philetor brachypterus*. Seven of the nine nonvolant species are known or believed to be non-native, transported to the island by humans (Flannery and White 1991). New Ireland's bat fauna is strikingly diverse, while its nonvolant mammal fauna is extremely depauperate. Bats likely have a predominant role in the ecology of the island, particularly with respect to seed dispersal and pollination of the flora. Introduced predatory mammals (cats, dogs, pigs) are probably altering the ecology. Although New Ireland's mammal fauna is still incompletely inventoried, the Weitin Valley appears to conserve much of the island's biodiversity, and it is an important reservoir of the native flora and fauna used as resources by local people.

Introduction

From the 1870s to the present, the islands of the Bismarck Archipelago, particularly New Britain, have been the source of occasional mammal specimens, as well as the focus of several scientific expeditions during which mammals were collected specifically or incidentally (reviewed by Koopman, 1979). New Ireland has been the focus of only two major mammal-collecting efforts: a bat-survey led by James Dale Smith in 1979, and a combined archaeological and mammal inventory study by a team from the Australian Museum in the late 1980s (Flannery and White 1991). Both of these expeditions were mainly restricted to the lowlands of the northern parts of the island, most of which is disturbed or secondary forest. Koopman (1979) published a mammal list for the Bismarck Archipelago with a discussion of the biogeography, and more recently Flannery and White (1991) reviewed the mammal fauna of New Ireland, including a review of the prehistoric and modern fauna in relation to the archaeological record of human occupation.

Prior to our expedition, the known mammal fauna of New Ireland comprised 35 species, including three marsupials, 29 bats, and three rats, as well as two introduced European species and an endemic rat known only from fossils (Flannery and White 1991). Archaeological evidence led Flannery and White (1991) to propose that all of the marsupials and two of the rats were introduced to New Ireland by humans, so that two rats (one extinct) were the only native nonvolant land mammals of the island. The complete list of mammals known from New Ireland is found in Appendix 7.

Although both birds and plants had been sampled in the montane habitat of southern New Ireland, we are not aware of any efforts to survey the montane mammal fauna prior to our expedition.

Methods

We inventoried mammals from four camps, two in the Weitin River Valley bottom, and two (1200 m, 1800 m) on the slopes of the highest massif of the Hans Meyer Range bordering the eastern side of the Weitin Valley (see Map and Gazetteer). Our mammal survey on New Ireland generally followed standard inventory methods. For capturing rodents we used break-back rat traps, Sherman folding traps, and a few Tomahawk traps. Only the break-back traps were successful. Hordes of ants and crickets in the lowlands and crickets in the higher camps stripped traps of peanut-butter based bait at night, and we found that toasted fresh coconut was far more effective.

Bats were captured in mist nets, chiefly by leaving large arrays of 20–30 nets (set for birds) open during the night, and also by setting some nets specifically for bats. At each of the four camps, one or two "canopy" nets with vertical stacks of 2–4 nets, reaching to heights of 10–20 m, were established in flyways between trees.

Vouchers or small series of specimens were preserved as either museum skins and skulls/skeletons, or fixed in formalin. Tissue samples of each species were preserved in buffer.

Results

During the expedition we recorded 23 species of native mammals, including four new bat records for the Island (Appendix 7). We also observed three species of feral domestic mammals (pig, dog, cat).

Terrestrial mammals—Our ground total trapping effort included 3,160 trap/nights, yielding 44 rodents. The four camps showed differences in the distribution of mammals. The largest native mammal and only macropod, the dusky pademelon (*Thylogale brunii*), was present at all elevations, and tracks and a sighting showed it to occur to above 2,000 m at the top of the massif. Both this species and feral pigs (*Sus scrofa*) were extremely numerous in the valley bottom and along the road, which implies that hunting pressure was low. Spotted cuscus (*Spilocuscus maculatus*) were also common in the valley bottom, but we saw none at higher elevations. At our lowest camp (Riverside), and at the two montane camps, rodent densities were equivalent, judging from trapping success rates (2%), but, curiously, we captured none at all at Weitin Base Camp (250 m), despite a large trapping effort (1005 trap/nights).

Our preliminary results indicate that *Rattus praetor* diminishes in numbers with increasing elevation: we caught four at 150 m, one at 1100 m, and none at 1800 m, but there was a smaller trapping effort at the highest site. The other two rodents were present from the lowest to the highest elevations, and we saw no evidence of any restricted montane fauna of non-flying mammals. There was an interesting change in the fur color of the arboreal rat *Melomys rufescens* along the altitudinal gradient, suggesting that either evolutionary differentiation is taking place, or that rat stocks introduced to the island at different times are partially segregated altitudinally. The *Melomys* we captured in the lowlands all had largely white bellies, those from 1800 m had gray bellies with white only at the throat, and those from 1100 m were intermediate.

Flannery and White (1991) described an endemic rat from New Ireland fossils, which occurred as recently as 3,000 years BP. We did not find evidence that this or any but the three known species currently inhabit Weitin watershed, but our sampling was limited to few sites and short periods, and we feel that more extensive and intensive trapping is needed before the presence of any other small terrestrial mammals on New Ireland can be ruled out.

Bats—In dramatic contrast to the terrestrial mammal fauna, the bat fauna of New Ireland is extraordinarily diverse: rich in both species and individuals. This oceanic island possesses 40% as many bat species as the subcontinental island of New Guinea. With its poor fauna of terrestrial mammals, bats are the dominant native mammals of New Ireland, and they surely must have had a predominant role in the past and present ecology of the island, particularly with respect to seed dispersal and pollination of the flora.

One of the factors that favors New Ireland as a habitat for bats is the presence of limestone caves. Previous collectors focused on caves and collected 13 species that we did not observe (we did not collect from caves), but most of these can also be expected to occur in the Weitin Valley if caves are present. The preservation of many of the bat species on New Ireland will depend on conserving its caves as well as its habitat diversity.

The echolocating bats (insectivorous Microchiroptera) of New Ireland are all widespread species known at least from New Guinea and other islands, but in many instances also from a much wider area of the Indopacific or Australia. We obtained two new records of vespertilionid bats for the Bismarcks, *Philetor brachypterus*, the range of which reaches India, and *Nyctophilus microtis*, for which this appears to be the first record outside of the island of New Guinea (Flannery 1990).

In addition, we found two nominal subspecies of the Emballonurid bat, Mosia *"nigrescens,"* in sympatry, and thus comprising two species, where only one was previously listed for New Ireland. We have not ascertained which of the two was the form previously reported from New Ireland by Smith and Hood (1981). We found *M. n. solomonis* in small groups (2–6) that were roosting beneath giant leaves of the arborescent stinging nettle (*Dendrocnide* sp.), but caught a *M. n. papuana* in a mist net.

In contrast to the small echlocating bats, the fruit bats (fruit and nectar feeding Megachiroptera) show a high degree of regional endemism, with five of the twelve species found on New Ireland being endemic to the Bismarcks. The most noteworthy new record we obtained on the expedition was of the flying fox *Pteropus gilliardorum* (formerly *P. gilliardi*, emended by Flannery 1995: 259). This bat was previously known only from the holotype, collected in 1958 at 1,600 m elevation in the Whiteman Range of New Britain (Van Deusen 1969). We obtained four specimens of *P. gilliardorum* at 1,800 m in forest that very closely corresponds in physiognomy with the description of the type locality on New Britain (Van Deusen 1969). The long dense fur of this species, and its absence from earlier surveys in the lowlands, lead us to believe that this beautiful bat, unlike any other mammal in the fauna, is a high-mountain habitat specialist. In contrast, the other Bismarck-endemic flying fox, *P. temmincki*, was captured in good numbers at all four camps, from 150 to 1,800 m (although we only collected vouchers at two camps). We captured *P. gilliardorum* by calling them down into the vicinity of a net placed in a large helipad clearing, by making chirps with a mechanical bird-call ("Audubon bird-call," which, incidentally, also brought in a *P. temmincki*). *Pteropus gilliardorum* was a strong-smelling bat that called loudly with bird-like chirping quite similar to that produced by the instrument.

Conservation Significance of The Weitin Valley

The Weitin River Valley and adjacent Hans Meyer Range are known to host 59% of all extant mammal species recognized from New Ireland, including all non-flying mammals, and all species endemic to the Bismarcks that occur on New Ireland. Most of the remaining 13 bat species are also likely to be present in the valley, especially if there are large caves (local people reported knowing of one). This region therefore shows excellent potential for conserving almost all of the mammal fauna of New Ireland.

In particular, the high elevation forests on each side of this valley are likely to be the only areas in New Ireland to contain its single montane mammal species, *Pteropus gilliardorum*. Because of the limited amount of upland habitat above 1000 m within the Bismarck Archipelago, this bat is potentially the species most vulnerable to extinction due to habitat conversion. Since it is thus far known only from five individuals from two localities (one on New Britain and one on New Ireland), it is impossible to assess its current status.

As far as can be judged from such a brief survey, the mammal populations appear to be in excellent condition, with little hunting pressure. However, it was disturbing to see large numbers of introduced feral vertebrate species, including dogs, cats, pigs, and cane toads. It may have been coincidental, but at our first camp (250 m), where we caught no rodents, we saw feral cats many times. These are likely to damage bird and reptile populations as well as those of small mammals, especially if the habitat shrinks or becomes increasingly disturbed.

Literature Cited

Andersen, K. 1912. Catalogue of the Chiroptera in the collection of the British Museum. British Museum of Natural History.

Flannery, T.F. 1990. The mammals of New Guinea. Robert Brown & Associates.

Flannery, T.F. 1995. Mammals of the South-west Pacific and Moluccan Islands. Comstock/Cornell University Press, Ithaca, NY.

Flannery, T.F. and J.P. White. 1991. Animal Translocation. National. Geogr. Res. and Explor., 7:96–113.

Griffiths, T.A., K.F. Koopman, and A. Starrett. 1991. The systematic relationship of *Emballonura nigrescens* to other species of *Emballonura* and to *Coleura* (Chiroptera: Emballonuridae). Amer. Mus. Novitates, No 2996.

Koopman, K.F. 1979. Zoogeography of mammals from islands off the northeastern coast of New Guinea. Amer. Mus. Novitates, No. 2690.

Smith, J.D. and C.S. Hood. 1981. Preliminary notes on bats from the Bismarck Archipelago (Mammalia: Chiroptera). Science in New Guinea, 8:81–121.

Smith, J.D. and C.S. Hood. 1983. A new species of tube-nosed fruit bat (*Nyctimene*) from the Bismarck Archipelago, Papua New Guinea. Occ. Pap. Mus. Texas Tech Univ., No 81.

Van Deusen, H.M. 1969. Results of the 1958-1959 Gilliard New Britain Expedition. 5. A new species of *Pteropus* (Mammalia, Pteropodidae) from New Britain, Bismarck Archipelago. American Museum Novitates 2731.

Gazetteer

Papua New Guinea: New Ireland

Weitin Base Camp (240–260 m):
14–24 January, 8–14 February, 1994
4° 30.2'S, 152° 56.3'E

Near junction of the Niagara and Weitin Rivers in the Weitin Valley. Lowland rainforest with scattered windthrow-induced gaps and natural disturbance due to periodic flooding of the area.

Closed canopy lowland rainforest with *Pometia pinnata* as the dominant canopy tree. *Neoscortechinia forbesii* was the frequency dominant among arborescent species. Understory open, with a herb layer consisting primarily of *Selaginella* cf. *durvillei, S. velutina,* rosette-stage *Calamus,* and various fern species. Lianes were physiognomically prominent, occurring as numerous ropes reaching into the canopy. Spiny-stemmed *Freycinetia* spp. were common in this latter group.

Lake Camp, Hans Meyer Range (1180–1200 m):
24 January–1 February, 1994
4° 27.2'S, 152° 56.3'E

Hans Meyer Range, NE of the Weitin River. Naturally forested ridge, surrounding a small natural lake. Pristine mossy forest, low montane by composition and physiognomy.

Closed-canopy lower montane cloud forest, with *Syzygium* spp. the stature-dominant overstory component. Among trees and shrubs, *Spathiostemon javensis* was the frequency dominant. There was substantial development of the bryophyte and epiphyte flora. Understories were typically open.

Top Camp, Hans Meyer Range (1800–1830 m):
1–7 February, 1994
4° 25.2'S, 152° 56.8'E

Hans Meyer Range, Angil Mountain. High elevation, undisturbed mossy forest, with considerable epiphytic development. Slippery clay substrate covered by bryophyte mats and ferny growth.

Closed canopy montane mossy forest, with *Metrosideros salomonensis* as the dominant overstory species. *Ascarina* sp. (*maheshwarii* or *philippinensis*) and *Platea excelsa* were frequency co-dominants in the subcanopy. The understory was composed primarily of herbaceous taxa, especially orchids and ferns, which collectively comprised more than half of the nontree species.

Riverside Camp, Weitin River (150-200 m):
3–16 February, 1994
4° 32'S, 152° 57'S

Approximately 25 km from Silur, 0.3 km from Weitin River, at the foothills of the Verron Range. Lowland site on the floor of the Weitin Valley along a small tributary to the Weitin River. Flat-bottomed, broad floodplain with unstable braided streams shifting over open beds of alluvial rock and gravel. Disturbed, previously logged lowland forest. Sampling was conducted in a remnant unlogged site within the overall logged area, in logged forest, along tributaries of the Weitin River, and in coastal habitats.

Various forest successional stages were present; the early stages dominated by *Casuarina equisetifolia* and the latter stages by *Pometia pinnata* in the canopy and *Dysoxylum* sp. and *Myristica* sp. in the understory.

Itinerary

Dates	Location	Georeference	Activities/Comments
10–12 January 1994	Port Moresby		organize team and supplies
13–14 January 1994	Port Moresby to New Ireland		in transit
14–24 January 1994 8–14 February 1994	Weitin Base Camp 260 m	4º 30.2'S 152º 56.3'E	survey and upland exploration
24 Jan.–1 February	Lake Camp 1200 m	4º 27.2'S 152º 56.3'E	survey
1–7 February 1994	Top Camp 1800 m	4º 25.2'S 152º 56.8'E	survey
3–16 February 1994	Riverside Camp 165 m	4º 32'S 152º 57'S	survey
16 February 1994	New Ireland to Rabaul to Port Moresby		in transit
16–18 February 1994	Port Moresby		debriefing

Appendices

Appendix 1

Plant Species recorded during the RAP Survey of Southern New Ireland

Wayne Takeuchi and Joseph Wiakabu

Key to coded entries:
Growth Status: C = climber, E = epiphyte, H = nonwoody terrestrial, T = fructicose or arborescent (tree).
Collection Site: C1 = Low Camp, C2 = Weitin Base Camp, C3 = Lake Camp, C4 = Top Camp.
Transect: I = species collection or record from inside the transect, O = species collection or record from outside the transect, NT = no transect established (Low Camp).
Voucher: HC = herbarium collection(s) made, LC = live collection made for Lae Botanical Garden, NC = no collection, sight record only.

Family	Species	Growth Status	Collection Site	Transect	Voucher
Cryptogams					
Bryidae					
	genus indet.	H	C3	O	HC
	Spiridens sp. (*aristifolius* or *reinwardtii)*	E	C3	O	HC
Marchantiales					
	cf. *Marchantia* sp.	E	C4	O	HC
	Marchantia sp.	H	C1	NT	HC
Ferns and Fern Allies					
Adiantaceae					
	Onychium siliculosum C. Chr.	H	C1	NT	HC
	Pityrogramma calomelanos (L.) Link	H	C1	NT	HC
	Syngramma schlechteri Brause	H	C2	O	HC
	Taenitis blechnoides (Willd.) Sw.	H	C3	O	HC
Aspleniaceae					
	Asplenium adiantoides (L.) C. Chr.	E	C2	O	HC
	Asplenium bipinnatifidum Bak.	E	C3	I	HC
	Asplenium caudatum Forst.	E	C3	I	HC
	Asplenium cuneatum Lam.	E	C2	I	HC
	Asplenium cymbifolium Christ	E	C4	I	HC
	Asplenium foersteri Brause	H	C3, C4	I	HC

continued

Family	Species	Growth Status	Collection Site	Transect	Voucher
	Asplenium keysserianum Ros.	E	C3	O	HC
	Asplenium cf. *laserpitiifolium* Lamk.	E	C2	I	HC
	Asplenium macrophyllum Sw.	E	C2, C3	I	HC
	Asplenium nidus L.	E	C2, C3	I	HC
	Asplenium pellucidum Lam.	E	C2	I	HC
	Asplenium phyllitidus Don ssp. *malesicum* Holtt.	E	C2	I	HC
	Asplenium cf. *schultzei* Brause	E	C4	I	HC
	Asplenium subemarginatum Ros.	H	C2	I	HC
	Asplenium tenerum Forst.	E	C2, C3	I	HC
	Asplenium sp. (approaching *affine*)	E	C3	O	HC
	Asplenium sp. (near *cuneatum* Lam.)	E	C3	O	HC
Athyriaceae					
	Callipteris prolifera (Lam.) Bory	H	C2	I	HC
	Diplazium bentamense Bl.	H	C3	O	HC
	Diplazium cordifolium Bl.	H	C3	O	HC
	Diplazium fuliginosum (Hook.) Price	H	C4	I	HC
	Diplazium (near) *simplicivenium* Holtt.	H	C4	I	HC
	Diplazium sp. A	H	C3, C4	I	HC
	Diplazium sp. B	H	C2	I	HC
Blechnaceae					
	Blechnum decorum Brause	H	C3	I	HC
	Blechnum ledermannii Brause	H	C4	I	HC
	Stenochlaena palustris (Burm. f.) Beddome	C	C2	I	HC
Cyatheaceae					
	Cyathea cf. *macgillivrayi* (Bak.) Domin.	T	C3	I	HC
	Cyathea moseleyi Bak.	T	C3	O	HC
	Cyathea cf. *moseleyi* Bak.	T	C2	I	HC
	Cyathea sp. A	T	C4	I	HC
	Cyathea sp. B	T	C4	I	HC
	Cyathea sp. C	T	C3	I	HC
	Cyathea sp. D	T	C3	I	HC
Davalliaceae					
	Davallia heterophylla Sm.	E	C2	I	HC
	Davallia pectinata Sm.	E	C2	I	HC
	Davallia repens (L.f.) Kuhn	E	C3	I	HC
	Davallia solida (G. Forst.) Sw.	E	C2	I	NC
	Leucostegia immersa (Wall.) Presl	E	C4	O	HC
	Leucostegia sp. (closer to *immersa* than to *pallida*)	E	C3	O	HC
Dennstaedtiaceae					
	Histiopteris incisa (Thunb.) J. Sm.	E	C4	I	HC
	Microlepia speluncae (L.) Moore	H	C2	I	HC
Dipteridaceae					
	Dipteris conjugata Reinw.	H	C4	O	NC
Dryopteridaceae					
	Acrophorus cf. *stipellatus* (Wallich.) Moore	H	C4	I	HC
	Arachnioides sp.	E	C4	I	HC

continued

Family	Species	Growth Status	Collection Site	Transect	Voucher
	Dryopteris cf. *subarborea* (Baker) C. Chr.	H	C4	I	HC
	Lastrea or *Dryopteris* sp.	H	C3	I	HC
	Polystichum cf. *aculeatum* (L.) Schott.	E	C3, C4	I	HC
Equisetaceae					
	Equisetum ramosissimum Desf. ssp. *debile* (Vauch.) Hauke	H	C2	O	HC
Gleicheniaceae					
	Gleichenia hirta Bl.	H	C3	I	HC
	Gleichenia milnei Baker	H	C3	I	HC
Grammitidaceae					
	Calymmodon clavifer (Hook.) Moore	E	C3, C4	I	HC
	Ctenopteris obliquata (Bl.) Copel.	E	C3	O	HC
	Ctenopteris subsecundodissecta (Zoll.) Copel.	E	C4	I	HC
	Ctenopteris sp.	E	C3	I	HC
	Grammitis dolichosora (Copel.) Copel.	E	C4	I	HC
	Grammitis reinwardtii Bl.	E	C3	I	HC
	Loxogramme scolopendrina (Bory) Presl	E	C2	I	HC
	Prosaptia contigua (Forst.) Presl	E	C3	I	HC
	Prosaptia rosenstockii Copel.	E	C3, C4	I	HC
	Scleroglossum pusillum (Bl.) v.A.v.R.	E	C3	O	HC
Hymenophyllaceae					
	Callistopteris apiifolia (Presl) Copel.	H	C3, C4	I	HC
	Cephalomanes boryanum (Kunze) Copel.	H	C2	I	HC
	Hymenophyllum angulosum Christ	E	C3, C4	I	HC
	Hymenophyllum cf. *imbricatum* (Bl.) Copel.	E	C4	I	HC
	Hymenophyllum cf. *rubellum* Ros.	E	C3	I	HC
	Macroglena meifolia Copel.	E	C3, C4	I	HC
	Pleuromanes pallidum (Bl.) Presl	E	C3, C4	I	HC
	Reediella humilis (Forst.) Picchi Sermolli	E	C2	I	HC
	Trichomanes aphlebioides Chr.	E	C2	I	NC
Lindsaea group					
	Lindsaea obtuse J. Sm.	E	C2, C4	I	HC
	Lindsaea pulchella (J. Sm.) Mett. ex Kuhn var. *blanda* (Mett. ex Kuhn) Kramer	E	C4	I	HC
	Lindsaea pulchella (J. Sm.) Mett. ex Kuhn var. *falcata* (Brause) Kramer	E	C3, C4	I	HC
	Lindsaea repens (Bory) Thwaites var. *delicatula* (Christ) Kramer	E	C3	I	HC
	Lindsaea tenuifolia Bl.	H	C2	O	HC
	Sphenomeris retusa (Cav.) Maxon	H	C2	O	HC
	Tapeinidium cf. *denhamii* (Hooker) C. Chr.	H	C3	I	HC
	Tapeinidium novoguineense Kramer	H	C4	I	HC
Lomariopsidaceae					
	Bolbitis quoyana (Gaud.) Ching	H	C2	I	HC
	Elaphoglossum novoguineense Ros.	E	C3	O	HC
	Elaphoglossum sp. (nearest *novoguineense* Ros.)	E	C4	I	HC
	Elaphoglossum ? *sordidum* Christ	E	C3	I	HC

continued

Family	Species	Growth Status	Collection Site	Transect	Voucher
	Elaphoglossum sp. A	E	C4	I	HC
	Elaphoglossum sp. B	E	C3	I	HC
	Lomagramma sinuata C. Chr.	C	C2	I	HC
	Lomariopsis intermedia (Copel.) Holtt.	C	C3	I	HC
	Teratophyllum articulatum (J. Sm.) Mett.	C	C3	O	HC
Lycopodiaceae					
	Huperzia nummulariifolia (Bl.) Jermy	E	C4	O	NC
	Huperzia phlegmaria (L.) Rothm.	E	C2, C3	I	HC
	Lycopodium clavatum L	H	C4	O	NC
	Lycopodium sp.	E	C4	O	HC
	Palhinhaea cernua (L.) Vasc. & Franco	H	C4	O	NC
Marattiaceae					
	Angiopteris evecta (Forst.) Hoff.	T	C2	O	NC
Oleandraceae					
	Nephrolepis biserrata (Sw.) Schott	H	C2	I	HC
	Nephrolepis hirsutula (Forst.) Presl	H	C2	I	HC
	Nephrolepis (near *hirsutula*)	H	C2	O	HC
	Nephrolepis cf. *lauterbachii* Christ	E	C3	I	HC
	Oleandra cuspidata Bak.	C	C3	O	HC
	Oleandra werneri Ros.	C	C3	O	HC
Ophioglossaceae					
	Ophioglossum pendulum L.	E	C2	I	HC
Osmundaceae					
	Leptopteris alpina (Bak.) C. Chr.	H	C4	I	HC
Polypodiaceae					
	Aglaomorpha drynarioides (Hook.) Roos	E	C2	I	HC
	Aglaomorpha heraclea (Kunze) Copel.	E	C3	I	HC
	Belvisia mucronata (Fee) Copel. var. *mucronata*	E	C3	O	HC
	Belvisia novoguineensis (Ros.) Copel.	E	C4	O	HC
	Belvisia spicata (L. f.) Mirbel ex Copel.	E	C4	O	HC
	Drynaria rigidula (Sw.) Beddome	E	C2	I	NC
	Drynaria sparsisora (Desv.) T. Moore	E	C1	NT	HC
	Goniophlebium demersum (Brause) Rodl-Linder	E	C3	O	HC
	Goniophlebium persicifolium (Desvaux) Beddome	E	C3	O	HC
	Goniophlebium serratifolium Brack.	E	C3	I	HC
	Microsorum linguiforme (Mett.) Copel.	E	C2	I	HC
	Microsorum membranifolium (R. Br.) Ching	H	C2	I	HC
	Microsorum powellii (Hook. & Bak.) Copel.	E	C3	I	HC
	Microsorum punctatum (L.) Copel.	E	C2	O	HC
	Microsorum punctatum (L.) Copel. (*glossophyllum facies*)	E	C2	I	HC
	Pyrrosia fallax (Alderw.) Price	E	C2	O	NC
	Pyrrosia longifolia (Burm. f.) Morton	E	C1	NT	HC
	Selliguea albidosquamata (Bl.) Parris	E	C4	O	HC
	Selliguea costulata (Cesati) Wagn. & Greth.	E	C3, C4	I	HC
	Selliguea enervis (Cav.) Ching	E	C3, C4	I	HC

continued

Family	Species	Growth Status	Collection Site	Transect	Voucher
	Selliguea cf. *enervis* (Cav.) Ching	E	C3	I	HC
	Selliguea sp.	E	C3	I	HC
Psilotaceae					
	Psilotum complanatum Sw.	E	C4	I	HC
	Psilotum nudum (L.) Griseb.	E	C2	O	HC
Pteridaceae					
	Pteris blumeana Agardh	H	C4	I	HC
	Pteris gardneri (Fee) Hooker	H	C2	I	HC
	Pteris ligulata Gaud.	H	C1	NT	HC
	Pteris moluccana Bl.	H	C2	O	HC
Schizaeaceae					
	Schizaea dichotoma (L.) Sm.	H	C2	O	NC
Selaginellaceae					
	Selaginella cf. *angustiramea* Muell. & Baker	E	C3, C4	I	HC
	Selaginella cf. *durvillei* R. Br.	H	C2	I	HC
	Selaginella velutina Cesati	H	C2	I	HC
	Selaginella sp.	C	C4	O	HC
Tectaria group					
	Cyclopeltis novoguineensis Ros.	H	C2	I	HC
	Pleocnemia dahlii (Hieron.) Holtt.	H	C2	I	HC
	Pleocnemia macrodonta (Fee) Holtt.	H	C2	I	HC
	Tectaria crenata Cav.	H	C2	I	HC
	Tectaria pleiosora (v.A.v.R.) C. Chr.	H	C2	I	HC
	Tectaria sp., cf. *crenata* Cav. or cf. *repanda* (Willd.) Holtt.	H	C2	O	HC
	Tectaria sp. ? nov. (section *Sagenia*)	H	C2	I	HC
Thelypteridaceae					
	Amphineuron pseudostenobasis (Copel.) Holtt.	H	C2	I	HC
	Coryphopteris cf. *pubirachis* (Bak.) Holtt.	H	C4	I	HC
	Metathelypteris polypodioides (Hook.) Holtt.	H	C2	O	HC
	Pronephrium cf. *beccarianum* (Cesati) Holtt.	H	C3	O	HC
	Sphaerostephanos hastatopinnatus (Brause) Holtt. var. ? nov.	H	C2	I	HC
	Sphaerostephanos unitus (L.) Holtt. var. *mucronatus* (Christ) Holtt.	H	C2	O	HC
	genus indet. (*Christella* or *Sphaerostephanos* sp.)	H	C2	I	HC
Vittariaceae					
	Antrophyum alatum Brack.	E	C2, C3	I	HC
	Vaginularia paradoxa (Fee) Mett.	E	C2	I	HC
	Vittaria elongata J. Sm. var. *elongata*	E	C2	I	HC
	Vittaria ensiformis Sw. var. *ensiformis*	E	C3	I	HC
	Vittaria scolopendrina (Bory) Thw.	E	C2	I	HC
	Vittaria sp.	E	C3	I	HC
Gymnosperms					
Gnetaceae					
	Gnetum latifolium Bl. var. *atifolium*	C	C2	O	HC

continued

Family	Species	Growth Status	Collection Site	Transect	Voucher
Podocarpaceae					
	Dacrycarpus aff. *imbricatus* (Bl.) de Laub.	T	C4	I	HC
	Falcatifolium aff. *papuanum* de Laub.	T	C3	I	HC
	Podocarpus section *Foliolatus* (between *P. rubens* and *P. insularis*)	T	C4	I	HC
	Prumnopitys amara (Bl.) de Laub.	T	C3	O	NC
Monocots					
Araceae					
	Epipremnum pinnatum (L.) Engl.	C	C2	I	NC
	Holochlamys beccarii Engl.	H	C2	I	HC
	Homalomena melanesica A. Hay	H	C2	I	HC
	Pothos cf. *rumphii* (Presl) Schott	C	C2	I	NC
	Rhaphidophora peekelii Engl. & Krause	C	C2	I	HC
	Rhaphidophora or *Scindapsus* sp.	C	C2, C3	I	HC
	Schismatoglottis sp.	H	C2	O	HC
Arecaceae					
	Areca novohibernica Becc.	H	C2	I	HC
	Calamus cf. *hollrungii* Becc.	C	C2	I	NC
	Calamus sp. (cf. *ralumensis* Warb.)	C	C2	I	NC
	Calyptrocalyx sp.	H	C3	I	HC
	Clinostigma collegarum Dransfield	T	C3	I	HC
	Hydriastele kaesesa (Laut.) Burret	T	C2	I	HC
	Ptychosperma sp. nov.	T	C2	I	HC
Commelinaceae					
	Pollia cf. *secundiflora* (Bl.) Back.	H	C2	I	HC
Corsiaceae					
	Corsia purpurata L.O. Williams var. *wiakabui* Takeuchi & Pipoly	H	C3	I	HC
	Corsia pyramidata van Royen	H	C4	I	HC
Cyperaceae					
	Cyperus cyperinus (Retz.) Valck. var. *cyperinus*	H	C2	O	HC
	Cyperus diffusus Vahl	H	C1	NT	HC
	Cyperus kyllingia Endl.	H	C1	NT	HC
	Gahnia sp. (aff. *javanica* Zoll. & Mor.)	H	C4	O	HC
	Mapania moseleyi Clarke	H	C3	I	HC
	Mapania palustris (Hassk ex Steud.) F.-Vill.	H	C3	I	HC
	Scleria scrobiculata Nees & Mey. ex Nees in Wight ssp. *scrobiculata*	H	C1	NT	HC
	genus indet.	H	C4	O	HC
Flagellariaceae					
	Flagellaria gigantea Hook. f.	C	C2	I	NC
Heliconiaceae					
	Heliconia indica Lam. var. *micholitzii* (Ridley) Kress	H	C2	O	HC
Liliaceae					
	Dianella sp.	H	C2	O	NC
Marantaceae					
	Donax cannaeformis (Forst. f.) K. Schum.	H	C2	O	NC

continued

Family	Species	Growth Status	Collection Site	Transect	Voucher
Orchidaceae (dets. by Howcroft)					
	Acriopsis javanica Reinw. ex Bl. var. *javanica*	E	C2	O	HC
	Agrostophyllum sp. (sect. *Dolichodesme*)	E	C2	I	LC
	Agrostophyllum sp.	E	C3	I	HC
	Appendicula sp.	E	C4	I	HC
	Arachnis beccarii Reichb. f. var. *beccarii*	E	C2	O	HC
	Bulbophyllum aff. *micholitzii* Rolfe	E	C2	I	HC
	Bulbophyllum musicola Schltr.	E	C4	O	HC
	Bulbophyllum pachyglossum Schltr.	E	C4	O	HC
	Bulbophyllum sp. A	E	C4	I	HC
	Bulbophyllum sp. B (section *Micromonanthe* or sect. *Schyphosepalum*)	E	C4	I	HC
	Bulbophyllum sp. C	E	C4	I	HC
	Bulbophyllum sp. D	E	C4	O	HC
	Cadetia echinocarpa Schltr.	E	C2	I	HC
	Calanthe triplicata (Willem) Ames	H	C2	I	HC
	Coelogyne asperata Lindl.	H	C2	I	LC
	Corybas epiphyticus (J. J. Sm.) Schltr.	E	C4	I	HC
	Dendrobium antennatum Lindley	E	C2	O	HC
	Dendrobium bracteosum Rchb. f.	E	C2	O	HC
	Dendrobium cuthbertsonii F.v.M.	E	C4	I	HC
	Dendrobium erosum (Bl.) Lindley	E	C3	I	HC
	Dendrobium cf. *fantasticum* L.O. Williams	E	C4	O	HC
	Dendrobium masarangense Schltr. ssp. *masarangense*	E	C3	I	HC
	Dendrobium rhodostictum F.v.M. & Krzl.	E	C2, C3	O	HC
	Dendrobium vexillarius J. J. Sm. ssp. *hansmeyerense* Howcroft & Takeuchi ined.	E	C4	I	HC
	Dendrobium sp. (section *Grastidium*)	E	C4	I	HC
	Diplocaulobium aff. *aureum*	E	C4	I	LC
	Dipodium pictum (Lindley) Reichb. f.	C	C2	O	HC
	Epiblastus sp.	E	C4	I	HC & LC
	Glossorhyncha sp.	E	C3	I	HC
	Goodyera grandis (Bl.) Bl.	H	C2	I	HC
	Liparis condylobulbon Rchb. f.	E	C2	I	HC
	Liparis cf. *gibbosa* Finnet	E	C3	I	HC
	Malaxis aff. *xanthochila* (Schltr.) Ames & C. Schweinf.	H	C2	I	HC
	Octarrhena angraecoides Schltr.	E	C3	I	HC
	Pedilochilus ciliolatum Schltr.	E	C4	I	HC
	Phaius terrestre (L.) Ormd.	H	C2	I	HC
	Phreatia elongata Schltr.	E	C4	O	HC
	Phreatia cf. *hypsorhynchos* Engler	E	C4	I	HC
	Phreatia loriae Schltr.	E	C2	I	HC
	Phreatia cf. *matthewsii* Reichb. f.	E	C2	I	HC
	Phreatia sp. A	E	C4	O	HC
	Phreatia sp. B	E	C3	O	HC
	Phreatia sp. C	E	C2	I	HC

continued

Family	Species	Growth Status	Collection Site	Transect	Voucher
	Pristiglottis cf. *longifolia* (Reichb. f.) Kores	E	C4	I	HC
	Rhynchophreatia aff. *micrantha*	E	C2, C3	I	HC
	Saccoglossum takeuchii Howcroft	E	C3	O	HC
	Spathoglottis plicata Bl.	H	C2	O	HC
	Taeniophyllum sp.	E	C2	I	HC
	Thrixspermum aff. *platystachys* (Bailey) Schltr.	E	C2	O	HC
	Trichotosia sp.	E	C3	I	HC
	genus indet., high canopy, pendant, several m long	E	C2	I	NC
	genus indet., vandaceous epiphyte	E	C2	I	NC
Pandanaceae					
	Freycinetia falcata Huynh	C	C3, C4	I	HC
	Freycinetia sp. A	C	C4	I	NC
	Freycinetia sp. B	C	C3	I	HC
	Freycinetia sp. C	C	C3	I	HC
	Pandanus sp. A	T	C2	I	NC
	Pandanus sp. B	T	C2	I	NC
Philesiaceae					
	Geitonoplesium cymosum Cunn.	C	C1	NT	HC
Poaceae					
	Bambusa vulgaris Schrad.	T	C1	NT	HC
	Centotheca lappacea (L.) Desv.	H	C1	NT	HC
	Eragrostis sp.	H	C1	NT	HC
	Leptaspis urceolata (Roxb.) R. Br.	H	C2	I	HC
	Thysanolaena maxima (Roxb.) O.K.	H	C2	O	HC
	genus indet. (Bambusoideae)	H	C3	I	HC
Zingiberaceae					
	Alpinia novae-*hibernia* Burtt & Smith	H	C3, C4	I	HC
	Alpinia novae-*pommeraniae* K. Schum.	H	C2	O	HC
	Alpinia oceanica Burk.	H	C2	I	HC
	Alpinia sp.	H	C4	I	HC
	Riedelia sp.	E	C3	I	HC
Dicots					
Acanthaceae					
	Lepidagathis sp.	H	C1	NT	HC
Actinidiaceae					
	Saurauia conferta Warb.	T	C3	O	HC
	Saurauia cf. *purgans* B.L. Burtt or aff.	T	C3	I	HC
Aizoaceae					
	Mollugo pentaphylla L.	H	C1	NT	HC
Anacardiaceae					
	Mangifera minor Bl.	T	C2	I	NC
	Rhus sp.	T	C2	O	NC
	Semecarpus sp. (either *lamii* Slis, or *papuanus* Laut.)	T	C2	I	HC

continued

Family	Species	Growth Status	Collection Site	Transect	Voucher
Annonaceae					
	Goniothalamus sp.	T	C2	O	HC
	Polyalthia sp.	T	C2	I	HC
	genus indet. (sterile)	T	C2	O	HC
Apocynaceae					
	Cerbera manghas L.	T	C2	O	HC
	Ichnocarpus frutescens (L.) W.T. Aiton	C	C1	NT	HC
	Parsonsia longiloba D.J. Middleton	C	C4	I	HC
Aquifoliaceae					
	Ilex cf. *versteeghii* Merr. & Perry	T	C4	I	HC
	Ilex sp.	T	C4	I	HC
Araliaceae					
	Arthrophyllum pacificum Philipson	T	C3, C4	I	HC
	Osmoxylon novoguineense (Scheff.) Becc.	T	C2	I	HC
	Osmoxylon pfeillii (Warb.) Philipson	T	C2	I	NC
	Plerandra stahliana Warb.	T	C2	I	HC
	Polyscias cumingiana (Presl) F.-Vill.	T	C2	I	HC
	Polyscias sp.	T	C3	I	HC
	Schefflera sp. (high canopy epiphyte)	E	C3	I	NC
Aristolochiaceae					
	Aristolochia sp.	C	C2	I	NC
Asclepiadaceae (Hoya dets. by P.I. Forster)					
	Dischidia sp.	E	C2	I	HC
	Hoya *australis* R. Br. ex Traill ssp. I (K. Hill) P.I. Forster & Liddle	C	C2	I	HC
	Hoya samoensis Seem.	C	C2	I	HC
	Hoya sp.	C	C2	O	HC
	Marsdenia sp. A	C	C3	I	HC
	Marsdenia sp. B	C	C3	I	HC
Asteraceae					
	Adenostemma lavenia (L.) Kuntze	H	C3	O	HC
	Blumea arfakiana Mart.	H	C2	O	HC
	Wedelia biflora (L.) DC.	H	C2	O	HC
Balsaminaceae					
	Impatiens sp.	H	C2	O	HC
Barringtoniaceae					
	Barringtonia sp.	T	C2	I	HC
Begoniaceae					
	Begonia sp.	H	C1	NT	HC
Bignoniaceae					
	Tecomanthe dendrophila (Bl.) K. Schum. & Laut.	C	C3	O	HC
Buddleiaceae					
	Buddleia asiatica Lour.	T	C1	NT	HC
Burseraceae					
	Canarium acutifolium (DC.) Merr.	T	C2	O	HC
	Canarium indicum L.	T	C2	I	HC

continued

Family	Species	Growth Status	Collection Site	Transect	Voucher
Caesalpiniaceae					
	Maniltoa cf. *schefferi* K. Schum. & Hollrung	T	C2	I	HC
Casuarinaceae					
	Casuarina equisetifolia Forst. & Forst. f.	T	C2	O	HC
Cecropiaceae					
	Poikilospermum sp.	C	C2	I	NC
Chloranthaceae					
	Ascarina sp. (*maheshwarii* or *philippinensis*)	T	C3, C4	I	HC
	Chloranthus erectus (Buch.-Ham.) Verdcourt	H	C4	I	HC
Clusiaceae (Calophyllum dets. by Stevens)					
	Calophyllum peekelii Laut.	T	C2	O	HC
	Calophyllum soulattri Burm.	T	C3	O	HC
	Calophyllum vexans Stevens	T	C2, C3	I, O	HC
	Calophyllum cf. *vexans* Stevens	T	C3	O	HC
	Garcinia sp.	T	C3	I	HC
Convolvulaceae					
	Merremia pacifica van Oostroom	H	C1	NT	HC
	Merremia sp.	C	C2	I	NC
Cucurbitaceae					
	Diplocyclos palmatus (L.) C. Jeffrey	C	C2	O	HC
Cunoniaceae					
	Caldcluvia celebica (Bl.) Hoogl.	T	C3, C4	O	HC
	Weinmannia fraxinea (D. Don) Miq.	T	C3, C4	I	HC
Daphniphyllaceae					
	Daphniphyllum gracile Gage	T	C4	I	HC
Datiscaceae					
	Octomeles sumatrana Miq.	T	C2	O	NC
Dilleniaceae					
	Tetracera nordtiana F.v var. *moluccana* (Martelli) Hoogl.	C	C2	O	HC
Ebenaceae					
	Diospyros ferrea (Willd.) Bakh., sens. lat.	T	C2	O	NC
Elaeocarpaceae					
	Aceratium oppositifolium DC.	T	C2, C3	I	HC
	Elaeocarpus culminicola Warb.	T	C4	O	HC
	Elaeocarpus sp.	T	C3	I	HC
	Sloanea sp.	T	C2	I	HC
	genus indet. (sterile)	T	C4	I	HC
Ericaceae (Dimorphanthera det. by Stevens)					
	Dimorphanthera angiliensis P.F. Stevens ined.	E	C3, C4	I	HC
	Rhododendron culminicolum F.v.M. var. *culminicolum*	E	C3, C4	I	HC
	Rhododendron superbum Sleum.	E	C4	I	HC
	Vaccinium sp.	E	C4	O	HC

continued

Family	Species	Growth Status	Collection Site	Transect	Voucher
Euphorbiaceae					
	Acalypha hellwigii Warb. var. *mollis* (Warb.) K. Schum. & Laut.	T	C2	O	HC
	Aporosa papuana Pax & Hoffm.	T	C3	I	HC
	Cleidion papuanum Laut.	T	C3	O	HC
	Endospermum (near) *myrmecophila* L.S. Sm.	T	C2	I	HC
	Glochidion sp. (between *grossum* and *nobile*)	T	C3	I	HC
	Glochidion sp. nov. (aff. *angulatum*)	T	C2	O	HC
	Glochidion sp. A	T	C3	I	HC
	Homalanthus longistylus Laut. & K. Schum.	T	C3	O	HC
	Macaranga dioica (Forst.) Muell. Arg.	T	C2, C3	O	HC
	Macaranga polyadenia Pax & Hoffm.	T	C3	I	HC
	Macaranga quadriglandulosa Warb. var. *quadriglandulosa*	T	C2	O	HC
	Macaranga sp. A	T	C3	I	HC
	Macaranga sp. B	T	C2	I	HC
	Mallotus echinatus Elm.	T	C1	NT	HC
	Mallotus mollissimus (Geisel.) Airy Shaw	T	C1	NT	HC
	Neoscortechinia forbesii (Hook. f.) Pax ex S. Moore	T	C2, C3	I	HC
	Pimelodendron amboinicum Hassk.	T	C2	I	HC
	genus indet. (sterile)	T	C2	I	HC
Fabaceae					
	Canavalia cf. *cathartica* Thouars	C	C2	O	HC
	Desmodium laxum DC.	T	C2	I	HC
	Mucuna novoguineensis Scheff.	C	C1	NT	HC
	Oxyrhynchus sp.	C	C2	O	HC
	Pterocarpus indicus Willd.	T	C2	O	NC
Flacourtiaceae					
	Casearia sp. (nearest *halmaherensis* Sloot.)	T	C2	O	HC
	Flacourtia cf. *rukam* Zoll. & Mor.	T	C2	I	HC
Gesneriaceae					
	Aeschynanthus kermesinus Schltr.	E	C3, C4	I	HC
	Aeschynanthus aff. *pachyanthus* Schltr.	E	C4	I	HC
	Aeschynanthus sp.	E	C3, C4	I	HC
	Agalmyla sp. (cf. *amabile*)	E	C3, C4	I	HC
	Cyrtandra erectiloba Gillett	T	C3	I	HC
	Cyrtandra fusco-vellea K. Schum.	E	C3	O	HC
	Cyrtandra laciniata Gillett	T	C3	O	HC
	Cyrtandra lutescens Gillett	T	C2	O	HC
	Cyrtandra sp. (section *Geodesme*)	E	C4	O	HC
	Cyrtandra sp. (section *Leucocyrtandra*)	T	C3	I	HC
	Cyrtandra sp. (section *Macrocyrtandra*)	T	C3	O	HC
Grossulariaceae					
	Polyosma cf. *integrifolia* Bl.	T	C3	I	HC
	Polyosma sp.	T	C4	I	HC
Halogoraceae					
	Gunnera macrophylla Bl.	H	C4	O	HC

continued

Family	Species	Growth Status	Collection Site	Transect	Voucher
Himantandraceae					
	Galbulimima belgraveana (F.v.M.) Sprague	T	C3	O	NC
Icacinaceae					
	Gomphandra montana (Schellenb.) Sleum.	T	C3	I	HC
	Platea excelsa Bl. var. *borneensis* (Heine) Sleum.	T	C3	I	HC
	Platea excelsa Bl. cf. var. *borneensis* (Heine) Sleum.	T	C4	O	HC
	Platea excelsa Bl. var. *microphylla* (Sleum.) Sleum.	T	C3	I	HC
Lauraceae					
	Actinodaphne sp.	T	C3	O	HC
	Cryptocarya cf. *depressa* Warb., or *C. kamahar* Teschn.	T	C2	I	HC
	Cryptocarya cf. *multipaniculata* Teschn.	T	C2	I	HC
	Cryptocarya sp. A	T	C3	O	HC
	Cryptocarya sp. B	T	C4	I	HC
	Endiandra sp.	T	C3	I	HC
	Litsea sp. (near *dielsiana* Teschn.)	T	C2	O	HC
	Litsea cf. *firma* (Bl.) Hook. f.	T	C3	I	HC
	Litsea sp. (possibly *L. firma*)	T	C3	O	HC
	Litsea cf. *solomonensis* Allen	T	C2	I	HC
	Litsea sp. A	T	C3	I	HC
	Litsea sp. B	T	C2	I	HC
	Litsea sp. C	T	C3	I	HC
	Neolitsea cf. *brassii* Allen	T	C3	O	HC
Leeaceae					
	Leea indica (Burm. f.) Merr.	T	C2	I	NC
	Leea macropus K. Schum. & Laut.	T	C2	I	HC
	Leea sp.	T	C2	O	HC
Loganiaceae					
	Fagraea ceilanica Thunb.	E	C4	I	HC
	Neuburgia corynocarpa (A. Gray) Leenh.	T	C3	O	HC
	Neuburgia sp.	T	C2	I	HC
	Strychnos minor L.	C/T	C2	O	HC
Loranthaceae (dets. by Barlow)					
	Decaisnina hollrungii (K. Schum.) Barlow	E	C2	O	HC
	Dendrophthoe curvata (Bl.) Miq.	E	C3	I	HC
	Sogerianthe sogerensis (Moore) Danser	E	C2, C3	I/O	HC
Magnoliaceae					
	Elmerrillia tsiampaca (L.) Dandy ssp. *tsiampaca*	T	C2	O	NC
Melastomataceae					
	Astronia or *Astronidium* sp. (sterile)	T	C2	I	HC
	Astronidium basinervatum Maxw.	T	C3, C4	I	HC
	Astronidium cf. *basinervatum* Maxw.	T	C3	I	HC
	Medinilla albida Merr. & Perry	E	C3, C4	I	HC
	Medinilla cf. *crassinervis* Bl.	C	C2	I	HC

continued

Family	Species	Growth Status	Collection Site	Transect	Voucher
	Medinilla (near *peekelii* Mansf.)	E	C2	I	HC
	Medinilla sp. A	T	C4	I	HC
	Medinilla sp. B	C	C3	I	HC
Meliaceae					
	Aglaia sapindina (F.v.M.) Harms	T	C1	NT	HC
	Aglaia cf. *sapindina* (F.v.M.) Harms	T	C2	I	HC
	Aglaia subcuprea Merr. & Perry	T	C2	I	HC
	Aglaia subminutiflora C. DC.	T	C2	I	HC
	Aglaia sp.	T	C2	I	HC
	Aphanamixis polystachya (Wall.) R.N. Parker	T	C2	I	HC
	Chisocheton ceramicus (Miq.) C. DC.	T	C2	I	HC
	Dysoxylum sp. A	T	C2	I	HC
	Dysoxylum sp. B	T	C2	I	HC
	Vavaea amicorum Benth.	T	C2	O	HC
Menispermaceae					
	Pycnarrhena ozantha Diels	C	C2	O	HC
Mimosaceae					
	Archidendron sp.	T	C2	I	HC
	Entada phaseoloides (L.) Merr.	C	C2	I	HC
Monimiaceae					
	Palmeria arfakiana Becc.	C	C3, C4	I	HC
Moraceae					
	Artocarpus vriesianus Miq.	T	C2	O	HC
	Ficus arfakensis King	T	C3	I	HC
	Ficus erythrospermum Miq.	T	C3	I	HC
	Ficus tinctoria Forst. f.	T	C3	I	HC
	Ficus wassa Roxb.	T	C3	I	HC
	Ficus sp. A	C	C3	I	HC
	Ficus sp. B	T	C2	I	HC
Myristicaceae					
	Horsfieldia psilantha de Wilde	T	C2	I	HC
	Horsfieldia sp.	T	C2	I	HC
	Myristica hollrungii Warb.	T	C2	I	HC
	Myristica inutilis Rich. ex A. Gray	T	C2	O	HC
	Myristica sp.	T	C2	O	HC
Myrsinaceae					
	Ardisia sp.	T	C3	I	HC
	Ardisia sp. (?conspecific with 9503/9648)	T	C2	O	HC
	Myrsine leucantha (K. Schum.) Pipoly	T	C3, C4	I	HC
	Myrsine sp. = *Rapanea rawacensis* (A. DC.) Mez	T	C3	O	HC
	Tapeinosperma commutatum Sleum.	T	C2	O	HC
Myrtaceae					
	Metrosideros salomonensis C.T. White	T	C3, C4	I	HC
	Syzygium decipiens (Koords. & Val.) Merr. & Perry	T	C3	O	HC
	Syzygium effusum (A. Gray) C. Muell.	T	C3	I	HC
	Syzygium longipes (Diels) Merr. & Perry	T	C2	O	HC
	Syzygium cf. *longipes* (Diels) Merr. & Perry	T	C2	I	HC

continued

Family	Species	Growth Status	Collection Site	Transect	Voucher
	Syzygium malaccense (L.) Merr. & Perry	T	C2	I	HC
	Syzygium ovalifolium (Bl.) Merr. & Perry	T	C2	O	HC
	Syzygium sp. (possibly *effusum*)	T	C3	O	HC
	Syzygium sp. A	T	C3	I	HC
	Syzygium sp. B	T	C3	I	HC
	Syzygium sp. C	T	C2, C3	I	HC
	Syzygium sp. or *Xanthomyrtus* sp. (sterile)	T	C3	I	HC
	Xanthomyrtus cf. *schlechteri* Diels	T	C4	I	HC
Nyctaginaceae					
	Pisonia longirostris Teysm. & Binn.	T	C2	I	HC
Ochnaceae					
	Schuurmansia henningsii K. Schum.	T	C2, C3, C4	I	NC
Piperaceae					
	Peperomia cf. *tenuipila* DC.	E	C4	I	HC
	Peperomia sp.	E	C3	I	HC
	Piper macropiper Pennant	C	C3	I	HC
	Piper mestonii Bailey	C	C3	I	HC
	Piper rodatzii K. Schum. & Laut.	C	C2	I	HC
	Piper cf. *rodatzii* K. Schum. & Laut.	C	C2	I	HC
	Piper sp.	C	C4	I	HC
Proteaceae					
	Helicia sp.	T	C4	I	HC
Rhamnaceae					
	Alphitonia ferruginea Merr. & Perry	T	C4	O	HC
	Alphitonia cf. *macrocarpa* Mansf.	T	C2	I	NC
Rhizophoraceae					
	Gynotroches axillaris Bl.	T	C2, C3, C4	I	HC
Rosaceae					
	Prunus dolichobotrys (Laut. & K. Schum.) Kalkman	T	C3	I	HC
	Prunus cf. *schlechteri* Kalkman, (possibly *arborea*)	T	C2	O	HC
	Rubus moluccanus L.	C	C4	I	NC
	Rubus royenii Kalkman var. *hispidus* Kalkman	C	C4	I	HC
Rubiaceae					
	Amaracarpus sp.	T	C4	I	HC
	Canthium cf. *korrense* (Val.) Kanch.	T	C2	O	HC
	Hedyotis schlechteri (Val.) Merr. & Perry	T	C4	I	HC
	Lasianthus papuanus Wernh.	T	C3, C4	I	HC
	Myrmecodia tuberosa Jack, cf. entity *salomonensis*	E	C3	I	HC
	Myrmecodia sp. (sterile)	E	C2	O	HC
	Neonauclea hagenii (Laut. & K. Schum.) Merr.	T	C2	I	HC
	Neonauclea cf. *obversifolia* (Val.) Merr. & Perry	T	C2	I	HC
	Ophiorrhiza sp.	H	C3	I	HC
	Psychotria amplithyrsa Val.	C	C3, C4	I	HC
	Psychotria crassiramula Sohmer	T	C3	I	HC

continued

Family	Species	Growth Status	Collection Site	Transect	Voucher
	Psychotria damasiana Sohmer	T	C2	O	HC
	Psychotria cf. *leptothyrsa* Miq.	T	C3	O	HC
	Psychotria micrococca (Laut. & K. Schum.) Val.	T	C3	I	HC
	Psychotria osiana Takeuchi & Pipoly	T	C3	I	HC
	Psychotria sp. A (liane)	C	C3	O	HC
	Psychotria sp. B (nonclimbing)	T	C4	I	HC
	Timonius sp.	T	C2	I	HC
	Uncaria sp. (sterile)	C	C2	I	NC
Rutaceae					
	Acronychia sp.	T	C4	I	HC
	Melicope sp. A	T	C3, C4	I	HC
	Melicope sp. B	T	C3	I	HC
	Melicope sp. C	T	C4	I	HC
	Micromelum minutum (Forst. f.) W. & A.	T	C1	NT	HC
Sabiaceae					
	Sabia pauciflora Bl.	C	C3	I	HC
Santalaceae					
	Dendromyza cf. *reinwardtiana* (Bl.) Danser	E	C4	O	HC
Sapindaceae					
	Elattostachys sp.	T	C2	I	HC
	Guioa sp.	T	C2	I	HC
	Harpullia sp.	T	C2	O	HC
	Mischocarpus sp.	T	C2	I	HC
	Pometia pinnata Forst.	T	C2	O	NC
	genus indet.	T	C4	I	HC
Sapotaceae					
	Pouteria firma (Miq.) Baehni	T	C3	I	HC
	Pouteria cf. *firma* (Miq.) Baehni	T	C2	I	HC
	Pouteria cf. *linggensis* (Burck) Baehni	T	C3	I	HC
	indet. sp. A (sterile)	T	C4	I	HC
	indet. sp. B (sterile)	T	C3	I	HC
Saxifragaceae					
	Dichroa febrifuga Lour.	T	C2	I	HC
Solanaceae					
	Solanum torvum Sw.	T	C2	O	HC
Sphenostemonaceae					
	Sphenostemon papuanus (Laut.) Steen. & Erdtman	T	C3, C4	I	HC
Sterculiaceae					
	Abroma augusta (L.) Willd.	T	C2	O	NC
	Melochia odorata L. f.	T	C1	NT	HC
	Sterculia cf. *ampla* Baker	T	C2	I	HC
	Sterculia schumanniana (Laut.) Mildbr.	T	C2	O	HC
	Sterculia sp.	T	C2	I	HC
Symplocaceae					
	Symplocos cochinchinensis (Lour.) S. Moore ssp. *leptophylla* (Brand.) Noot.	T	C3, C4	I	HC

continued

Family	Species	Growth Status	Collection Site	Transect	Voucher
Theaceae					
	Eurya sp.	T	C3, C4	I	HC
	Gordonia amboinensis (Miq.) Merr.	T	C3	I	HC
	Ternstroemia cherryi (F.M. Bailey) Merr.	T	C2	O	HC
Trimeniaceae					
	Trimenia weinmanniifolia Seeman ssp. *bougainvilleensis* Rodenburg	T	C4	I	HC
Ulmaceae					
	Celtis latifolia (Bl.) Planch.	T	C2	I	HC
	Celtis cf. *philippensis* Blanco	T	C2	I	HC
	Celtis rigescens (Miq.) Planch.	T	C2	I	HC
	Gironniera celtidifolia Gaudich	T	C3	I	HC
	Parasponia rugosa Bl.	T	C2	O	HC
Urticaceae					
	Boehmeria platyphylla D. Don	T	C1	NT	HC
	Cypholophus sp.	T	C3	O	HC
	Dendrocnide cf. *latifolia* (Gaud.) Chew	T	C2	I	HC
	Dendrocnide cf. *longifolia* (Hemsl.) Chew	T	C2	I	HC
	Elatostema densum Winkler	E	C4	I	HC
	Elatostema novae-britanniae Laut.	E	C3	I	HC
	Elatostema (near) *novoguineensis* Winkler	H	C2	I	HC
	Elatostema cf. *polioneurum* Hall. f.	H	C3	O	HC
	Elatostema sesquifolium Hassk	H	C2, C3	I	HC
	Leucosyke capitellata (Poir) Chew	T	C1	NT	HC
	Nothocnide repanda (Bl.) Bl.	C	C2	I	HC
	Pipturus argenteus (Forst. f.) Wedd.	T	C2	I	HC
	Procris cf. *frutescens* Bl.	E	C2	I	HC
	Villebrunea trinervis Wedd.	T	C2	O	HC
	genus indet.	T	C2	O	HC
Verbenaceae					
	Clerodendrum buchanani (Roxb.) Walp.	T	C2	O	HC
	Clerodendrum buruanum Miq.	T	C3	O	HC
	Gmelina dalrympleana (F.v.M.) H.J. Lam.	T	C2	I	HC
	Premna serratifolia L.	T	C1	NT	HC
	Vitex cofassus Reinw. ex Bl.	T	C2	O	HC
Viscaceae					
	Korthalsella cf. *japonica* (Thunb.) Engl.	E	C3	O	HC
Vitaceae					
	Cayratia japonica (Thunb.) Gagn.	C	C1	NT	HC
	Cissus penninervis (F.v.M.) Planch.	C	C3	I	HC
Winteraceae					
	Zygogynum sp. (=cf. *Bubbia haplopus* B.L. Burtt)	T	C2, C3, C4	I	HC
family indet. (sterile)		T	C2	I	HC

continued

Appendix 2

List of Butterfly Species Encountered in Southern New Ireland, Mainly in the Weiten Valley

Larry J. Orsak, Nick Easen, and Tommy Kosi

*= Bismarck regional endemic; ** = island endemic

Papilionidae

1. *Ornithoptera priamus urvilliana* *
2. *Papilio ulysses ambiguus* *
3. *Papilio euchenor novohibernicus* **
4. *Papilio ambrax ambrax*
5. *Papilio fuscus cilix* *
6. *Papilio aegeus oritis* *
7. *Graphium macfarlaneii seminigra* *
8. *Graphium codrus segonax* *
9. *Graphium kosii* **?

Pieridae

10. *Elodina primulari primularis* *
11. *Delias narses* *
12. *Eurema hecabe kerawara* *

Danaidae

13. *Tellervo soilus ?sarcapus*
14. *Danaus hamata obscurata*
15. *Danaus juventa sobrinoides*
16. *Danaus melusine rotnndata* *
17. *Euploea eboraci* * (rare)
18. *Euploea stephensii bismarkiana* *

Nymphalidae

19. *Vindula arsinoe lemina* **
20. *Cethosia obscura obscura* *
21. *Cethosia chrysippe chrysippe*
22. *Precis villida*
23. *Hypolimnas ?pithoeka*
24. *Hypolimnas deois ?panopion*
25. *Cyrestis adaemon* *
26. *Cyrestis acilia fratercula* *
27. *Pantoporia venilia novohannoverana* *
28. *Pantoporia consimilis novahibernica* *
29. *Neptis nausicaa nausicaa*
30. *Prothoe ?ribbei* (closest to this species, but does not fit)

Satyridae

31. *Mycalesis terminus matho*
32. *Mycalesis durga* ssp.
33. *Melantis amabilis amabilis* *
34. *Elymnias vitellia*

Amathusiidae

35. *Taenaris phorcas phorcas*
36. *Taenaris myops* ssp?

Lycaenidae

37. *Arhopala kiriwini*
38. *Arhopala helianthus*
39. *Waigeum coruscans ?dinawa*
40. *Psychonotis browni*
41. *Prosotas nora nora*
42. *Catochrysops strabo* ssp?
43. *Catochrysops panormus papuana*
44. *Philiris tombara* *
45. *Eupsychellus dionicius*
46. *Celastrina philippina lychorida* *

Appendix 3

Amphibian and Reptile Species recorded during the RAP Survey of Southern New Ireland

Allen Allison and Ilaiah Bigilale

	Sampling Site				
	Coast	Weitin	Lake	Summit	Riverside
Elevation	Sea level	250 m	1180 m	1830 m	200 m
Sampling effort	Ad hoc	11 days	12 days	3 days	12 days
Species					
Amphibians: Frogs					
Bufonidae					
Bufo marinus	x	x	x	x	x
Hylidae					
Litoria infrafrenata	x	x	x		
Litoria thesaurensis		x	x		x
Ranidae					
Platymentis magnus			x		x
Platymentis papuensis		x	x		x
Platymentis schmidti		x	x		x
Platymantis browni		x	x		x
Reptiles: Lizards					
Agamidae					
Hypsilurus modestus		x	x		x
Reptiles: Geckos					
Gekkonidae					
Gehyra mutilata	x				
Gehyra oceanica	x				
Gekko vittatus	x				
Hemidactylus frenatus	x				
Nactus sp.		x			x

continued

	Sampling Site				
	Coast	Weitin	Lake	Summit	Riverside
Elevation	Sea level	250 m	1180 m	1830 m	200 m
Sampling effort	Ad hoc	11 days	12 days	3 days	12 days
Scincidae					
Carlia fusca		x	x		
Emoia atrocostata	x				
Emoia bismarckensis		x	x		
Emoia caeruleocauda		x	x		x
Emoia cyanogaster	x				
Emoia impar		x	x		x
Emoia jakati		x	x		
Eugongylus albofasciolatus					x
Lamprolepis smaragdina		x	x		
Sphenomorphus derooyae			x		
Sphenomorphus stickeli	x	x			x
Sphenomorphus cf. *jobiensis*	x	x			x
Sphenomorphus sp.					x
Varanidae					
Varanus indicus	x				
Reptiles: Snakes					
Pythonidae					
Bothrochilus boa	x	x			x
Candoia aspera	x				
Canoia carinata	x	x			x
Morelia amethistina	x	x			x
Colubridae					
Boiga irregularis	x	x	x		x
Dendrelaphis calligastra	x	x			x
Stegonotus heterurus		x	x		
Tropidonophis hypomelas					
Elapidae					
Aspidomorphus muelleri	x				
Typhlopidae					
Ramphotyphlops flaviventer		x			
Ramphotyphlops subocularis		x			
Typhlops depressiceps		x			
Total # Species per Site	**18**	**25**	**12**	**1**	**24**

Appendix 4

Land and Freshwater Birds of New Ireland

Bruce M. Beehler, J. Phillip Angle, David Gibbs, Michael Hedemark, and Daink Kuro

	Status on New Ireland (from Mayr and Diamond 2001)	Beehler, 1978	Finch & McKean,1987	Jones & Lambley, 1987	Gibbs, In litt.	RAP, 1994
Bird Species						
Tachybaptus ruficollis	resident					x
Egretta sacra	resident		x	x		
Ardeola striata	resident			x		
Nycticorax caledonicus	resident					x
Ixobrychus flavicollis	resident		x	x		x
Threskiornis spinicollis	vagrant					
Pandion haliaetus	resident			x	x	x
Aviceda subcristata	resident	x	x	x	x	x
Haliastur indus	resident	x	x	x	x	x
Haliaeetus leucogaster	resident		x	x		x
Accipiter meyerianus	resident?					x?
Accipiter brachyurus	resident		x?			x
Accipiter novaehollandiae	resident		x	x	x	x
Anas superciliosa	resident					
Megapodius eremita	resident	x	x		x	x
Coturnix chinensis	resident					
Gallirallus philippensis	resident					
Gymnocrex plumbeiventris	resident?					
Rallina tricolor	resident					
Porzana cinerea	resident					
Amaurornis olivaceus	resident					x
Esacus magnirostris	resident			x		x
Charadrius dubius	resident					
Tringa hypoleucos	northern migrant	x	x	x		x
Columba pallidiceps	resident		x			
Columba vitiensis	resident	x				

continued

	Status on New Ireland (from Mayr and Diamond 2001)	Beehler, 1978	Finch & McKean,1987	Jones & Lambley, 1987	Gibbs, In litt.	RAP, 1994
Macropygia amboinensis	resident	x	x	x	x	x
Macropygia nigrirostris	resident					x
Macropygia mackinlayi	?		x?			
Reinwardtoena browni	resident	x	x		x	x
Chalcophaps stephani	resident		x	x	x	x
Caloenas nicobarica	resident		x			
Gallicolumba beccarii	resident					x
Gallicolumba jobiensis	resident					
Ptilinopus superbus	resident				x	x
Ptilinopus insolitus	resident	x		x	x	x
Ptilinopus rivoli	resident	x	x	x	x	x
Ducula bicolor	resident				x	x
Ducula rubricera	resident	x	x	x	x	x
Ducula pistrinaria	resident		x			
Ducula melanochroa	resident	x	x	x	x	x
Ducula finschii	resident	x			x	
Gymnophaps albertisi	resident	x	x			x
Trichoglossus haematodus	resident	x	x	x	x	x
Lorius hypoinochrous	resident	x	x		x	x
Lorius albidinuchus	resident	x	x		x	x
Charmosyna rubrigularis	resident	x	x	x	x	x
Charmosyna placentis	resident	x	x	x	x	x
Micropsitta finschii	resident		x		x	x
Micropsitta bruijnii	resident	x	x		x	x
Geoffroyus heteroclitus	resident	x	x			x
Eclectus roratus	resident	x	x	x	x	x
Loriculus tener	resident		x			
Cuculus saturatus	northern migrant		x	x		
Cacomantis variolosus	resident		x		x	x
Chrysococcyx lucidus	Australian migrant		x			
Eudynamys scolopacea	resident	x	x		x	x
Scythrops novaehollandiae	Australian migrant?					
Centropus violaceus	resident	x	x		x	x
Centropus ateralbus	resident	x	x	x	x	x
Ninox variegata	resident	x	x	x	x	x
Podargus ocellatus	resident?				x?	
Caprimulgus macrurus	resident		x			x
Hemiprocne mystacea	resident	x	x	x	x	x
Collocalia esculenta	resident		x	x	x	x
Collocalia vanikorensis	resident	x	x	x	x	x
Collocalia spodiopygia	resident	x	x	x	x	x
Collocalia whiteheadi	resident					x
Halcyon sancta	Australian migrant		x			x
Halcyon chloris	resident		x	x	x	x
Halcyon saurophaga	resident	x		x		x
Alcedo atthis	resident		x	x	x	x
Alcedo pusilla	resident					

continued

	Status on New Ireland (from Mayr and Diamond 2001)	Beehler, 1978	Finch & McKean,1987	Jones & Lambley, 1987	Gibbs, In litt.	RAP, 1994
Ceyx lepidus	resident					x
Ceyx websteri	resident		x			x
Eurystomus orientalis	resident			x		x
Rhyticeros plicatus	resident	x	x	x	x	x
Pitta erythrogaster	resident		x		x	x
Hirundo tahitica	resident			x		x
Hirundo striolata	northern migrant				x	
Motacilla flava	northern migrant					
Coracina papuensis	resident	x	x	x	x	x
Coracina lineata	resident		x			x
Coracina tenuirostris	resident	x	x	x		x
Lalage leucomela	resident	x	x	x	x	x
Saxicola caprata	resident					
Turdus poliocephalus	resident	x				x
Acrocephalus australis	resident					
Megalurus timoriensis	resident					
Cisticola exilis	resident			x	x	x
Phylloscopus trivirgatus	resident	x	x	x	x	x
Rhipidura dahli	resident	x			x	x
Rhipidura rufiventris	resident	x	x	x	x	x
Rhipidura leucophrys	resident	x	x	x		x
Monarcha verticalis	resident	x	x		x	x
Monarcha chrysomela	resident	x	x	x	x	x
Myiagra alecto	resident	x	x	x	x	x
Myiagro hebetior	resident	x				x
Microeca sp.	?		?			
Pachycephala pectoralis	resident	x	x		x	x
Dicaeum eximium	resident	x	x	x	x	x
Nectarinia aspasia	resident	x	x	x	x	x
Nectarinia jugularis	resident		x	x	x	x
Zosterops hypoxantha	resident	x		x	x	x
Myzomela cruentata	resident	x	x	x	x	x
Myzomela pulchella	resident	x	x		x	x
Philemon eichhorni	resident	x	x		x	x
Eythrura trichroa	resident					x
Lonchura forbesi	resident		x	x	x	x
Lonchura hunsteini	resident					
Aplonis cantoroides	resident					x
Aplonis metallica	resident	x	x	x	x	x
Mino dumontii	resident	x	x	x	x	x
Dicrurus megarhynchus	resident	x	x	x	x	x
Artamus insignis	resident			x	x	x
Corvus orru/insularis	resident	x	x	x	x	x
Total	**119**	**53**	**73**	**53**	**61**	**86**

Appendix 5

General Bird Census Results from Southern New Ireland RAP

Bruce M. Beehler, J. Phillip Angle, Michael Hedemark, and Daink Kuro

Counts: number of days recorded/number of days the site was worked (in parentheses is high count of individuals) e.g., 1/9(4) indicates the species was recorded one of 9 days, with a high count of 4.

For Weitin and Coastal sites, simple presence/absence was recorded (e.g., x = present, recorded at site).

	Sites				
Camp	Coastal	Riverside	Weitin	Lake	Top
Elevation	sea level	150 m	240 m	1175 m	1840 m
# of days of survey	4 days	9 days	11 days	7 days	6 days
Bird Species Recorded					
Tachybaptus ruficollis				5/7(2)	
Fregata cf. *ariel*	x				
Nycticorax caledonicus			x		
Ixobrychus flavicollis		2/9			
Pandion haliaetus			x		
Aviceda subcristata		4/9(5)			
Haliastur indus		6/9	x		
Haliaeetus leucogaster	x		x	2/7(1)	2/6
Accipiter novaehollandiae		4/9	x	1/7	
Accipiter meyerianus					1/6
Accipiter brachyurus				3/7	1/6
Megapodius eremita		1/9			
Amaurornis olivaceus		2/9			
Esacus magnirostris	x				
Pluvialis fulva	x				
Numenius phaeopus	x				
Tringa hypoleucos	x	7/9			
Tringa brevipes	x				
Sterna bergii	x				

continued

	Sites				
Camp	Coastal	Riverside	Weitin	Lake	Top
Elevation	sea level	150 m	240 m	1175 m	1840 m
# of days of survey	4 days	9 days	11 days	7 days	6 days
Ptilinopus rivoli				4/7(5)	5/6
Ptilinopus superbus		6/9	x		
Ptilinopus insolitus		8/9(3)	x		
Ducula bicolor	x	4/9(6)			
Ducula rubricera		9/9(6)			
Ducula melanochroa		3/9(4)	x	5/7(4)	6/6
Macropygia amboinensis	x	8/9(2)	x	5/7(5)	4/6
Macropygia nigrirostris			x	3/7	6/6
Reinwardtoena browni		7/9(6)	x	6/7(3)	
Chalcophaps stephani		9/9	x		
Gymnophaps albertisi			x	2/7	2/6
Trichoglossus haematodus	x	9/9	x	7/7	6/6
Lorius hypoinochrous		5/9	x		
Lorius albidinuchus				7/7	
Charmosyna rubrigularis		6/9(30)	x	7/7(20+)	6/6
Charmosyna placentis		4/9	x		
Micropsitta finschii		5/9(5)			
Micropsitta bruijnii				1?	6/6
Geoffroyus heteroclitus					5/6
Eclectus roratus		9/9	x		
Cacomantis variolosus		6/9	x		
Eudynamys scolopacea		7/9		2/7	
Centropus violaceus		4/9			
Centropus ateralbus		8/9(4)	x	?1/7	
Ninox variegata		7/9	x	7/7	4/6
Caprimulgus macrurus		3/9(3)	x		
Collocalia esculenta		7/9			
Collocalia vanikorensis	x	8/9(5)	x	?3/7?	?
Collocalia spodiopygia		5/9		?	?
Collocalia whiteheadi		3/9			
Hemiprocne mystacea	x	3/9	x		
Alcedo atthis	x	1/9			
Ceyx lepidus			x	1/7	
Ceyx websteri		3/9	x		
Halcyon chloris		5/9	x	?1/7	2/6
Halcyon saurophaga	x	1/9			
Eurystomus orientalis		1/9			
Rhyticeros plicatus		9/9(8)	x		
Pitta erythrogaster		3/9	x		
Hirundo tahitica		4/9			
Coracina papuensis		5/9	x	1	4/6
Coracina tenuirostris			x	3/7	
Coracina lineata		2/9			
Lalage leucomela		9/9	x	7/7	
Turdus poliocephalus				?2/7	6/6

continued

	Sites				
Camp	Coastal	Riverside	Weitin	Lake	Top
Elevation	sea level	150 m	240 m	1175 m	1840 m
# of days of survey	4 days	9 days	11 days	7 days	6 days
Cisticola exilis		3/9			
Phylloscopus trivirgatus				3/7	6/6
Rhipidura dahli			x	5/7	
Rhipidura rufiventris		8/9	x	-	1/6
Rhipidura leucophrys		5/9			
Monarcha verticalis		4/9	x	1/7	
Monarcha chrysomela		8/9	x	?1/7	
Myiagra alecto		7/9	x		
Myiagro hebetior		5/9	x	2/7	
Pachycephala pectoralis		1/9	x	6/7(6)	4/6
Nectarinia aspasia		5/7(3)			
Nectarinia jugularis	x				
Myzomela cruentata		5/9	x	7/7	
Myzomela pulchella		8/9		1/7	6/6
Philemon eichhorni				7/7(4)	6/6
Dicaeum eximium		8/9	x	3/7	
Zosterops hypoxantha			x	6/7(10+)	6/6
Lonchura forbesi	x				
Eythrura trichroa	x	1/9		1/7	1/6
Aplonis cantoroides	x	1/9			
Aplonis metallica		9/9(25)	x		
Mino dumontii		9/9(4)		1/7	
Dicrurus megarhynchus		9/9	x	7/7	
Artamus insignis		1/9	x		
Corvus orru/insularis		7/9			

Appendix 6

Bird Census Results from Three RAP Camps

Bruce M. Beehler, J. Phillip Angle, Michael Hedemark, and Daink Kuro

Bird Species	Riverside Camp 165 meters		Lake Camp 1200 meters		Top Camp 1800 meters	
	Sum total individuals	# censuses observed	Sum total individuals.	# censuses observed	Sum total individuals	# censuses observed
Aviceda subcristata	2	1				
Haliastur indus						
Accipiter brachyurus			2	2		
Accipiter novaehollandiae	2	2				
Amaurornis olivaceus	3	2				
Macropygia amboinensis	16	7	18	7	2	1
Macropygia nigrirostris			1	1	19	7
Reinwardtoena browni	2	1	4	4		
Chalcophaps stephani	8	6				
Ptilinopus superbus	9	5				
Ptilinopus insolitus	18	6				
Ptilinopus rivoli			26	7	5	2
Ducula rubricera	13	7				
Ducula melanochroa	6	2	14	7	26	7
Gymnophaps albertisi			1	1	10	4
Trichoglossus haematodus	13	6	12	4	3	2
Lorius hypoinochrous	2	1				
Lorius albidinuchus			2	1		
Charmosyna rubrigularis	5	1	19	6	23	7
Charmosyna placentis	7	4				
Micropsitta finschii	7	6				
Micropsitta bruijnii					18	7
Eclectus roratus	4	3				
Cacomantis variolosus	6	5				
Centropus violaceus	6	3				
Centropus ateralbus	20	7	1?	1?		

continued

Bird Species	Riverside Camp 165 meters		Lake Camp 1200 meters		Top Camp 1800 meters	
	Sum total individuals	# censuses observed	Sum total individuals.	# censuses observed	Sum total individuals	# censuses observed
Hemiprocne mystacea	1	1				
Collocalia esculenta	2	2				
Collocalia vanikorensis	25	7				1
Collocalia spodiopygia	1	1				
Halcyon sancta						
Halcyon chloris	5	4			1	1
Ceyx lepidus	3	2				
Eurystomus orientalis	1	1				
Rhyticeros plicatus	8	5				
Pitta erythrogaster	1	1				
Coracina papuensis	2	1				
Coracina tenuirostris			3	3		
Lalage leucomela	17	7	5	5		
Turdus poliocephalus					18	6
Phylloscopus trivirgatus					21	6
Rhipidura dahli			4	3	1	1
Rhipidura rufiventris	10	7				
Monarcha verticalis	2	1				
Monarcha chrysomela	13	7				
Myiagra alecto	4	4				
Myiagro hebetior	3	2				
Pachycephala pectoralis			7	6	4	4
Dicaeum eximium	14	7	1	1		
Nectarinia aspasia	7	5				
Zosterops hypoxantha			12	2	20	6
Myzomela cruentata	7	4	25	7		
Myzomela pulchella					24	7
Philemon eichhorni			16	7	21	7
Aplonis metallica	46	7				
Mino dumontii	10	4				
Dicrurus megarhynchus	13	6	8	5		
Corvus orru/insularis	2	1				
Total number of individuals	**346**		**180**		**216**	
Number of species	**42 spp.**		**20 spp.**		**16 spp.**	

Appendix 7

Mammals Known from New Ireland

Louise H. Emmons and Felix Kinbag

Legend
Boldface = New record for New Ireland
* = specimen collected;
1 = previous expeditions, summarized by Flannery and White (1991)
2 = previous survey, Smith and Hood (1983)

We are grateful to the late Dr. Karl Koopman (American Museum of Natural History) for identification of some vespertilimid bats, and to Fiona Reid (Royal Ontario Museum) for helpful information about several bat species.

	Weitin Base Camp (260 m)	Lake Camp (1200 m)	Top Camp (1800 m)	Riverside Camp (165 m)	Previous Surveys
Marsupials					
*Spilocuscus maculatus**	x			x	1
Phalanger orientalis		x			1
*Thylogale brunii**	x	x	x	x	1
Bats					
Megachiroptera					
*Dobsonia moluccensis**		x			1
*Dobsonia praedatrix**	x				1
Nyctimene albiventer					2
Nyctimene major					1
*Nyctimene masalai**[1]	x	x		x	1
*Macroglossus minimus**	x	x			1
*Melonycteris melanops**	x	x			1
Pteropus gilliardorum*			x		
Pteropus neohibernicus	x			x	1
*Pteropus temmincki**	x	x	x	x	1
*Rousettus amplexicaudatus**				x	1
*Syconycteris australis**	x	x	x	x	1

continued

	Weitin Base Camp (260 m)	Lake Camp (1200 m)	Top Camp (1800 m)	Riverside Camp (165 m)	Previous Surveys
Microchiroptera					
Aselliscus tricuspidatus					1
Emballonura beccarii					1
Emballonura dianae					1
Emballonura furax					1
Emballonura raffrayana					1
Mosia nigrescens papuana*[2]	x				
*Mosia nigrescens solomonis**[2]				x	1?
*Hipposideros ater**	x				1
Hipposideros calcaratus					1
Hipposideros cervinus					1
*Hipposideros diadema**	x	x		x	1
Hipposideros maggietaylorae					1
Miniopterus macrocneme					1
Miniopterus propitristis					1
Miniopterus schreibersii					1
Nyctophilus microtis*	x	x			
Philetor brachypterus*	x				
*Pipistrellus angulatus**?	x	x			1
Pipistrellus papuanus					1
Rhinolophus megaphyllus					1
Rodents					
*Melomys rufescens**		x	x	x	1
*Rattus exulans**		x	x	x	1
*Rattus praetor**		x		x	1
Introduced/Human Commensals					
Canis familiaris	x				1
Felis sylvestris	x			x	1
Sus scrofa	x	x	x		2

Taxonomic notes:

[1]The definition of species in this genus in the Bismarcks is confused (Flannery and White 1991). Our specimens seem to correspond to *N. masalai* Smith and Hood 1983. We compared them to a series of *N. albiventer* from the Moluccas, adjacent to the type locality, and to *N. cephalotes* from the Celebes, also not far from its type locality. Our specimens do not appear to differ in many significant features from *N. cephalotes*, but differ radically in many features from *N. albiventer*, as shown and illustrated by Smith and Hood (1983). Flannery and White (1991) found that *Nyctimene* specimens which they collected on New Ireland were also close to *N. masalai*, but stated that these did not differ from *N. albiventer* collections "from throughout Melanesia." They conclude that "we find no basis for distinguishing *N. masalai* from our New Ireland sample that we find similar to *N. albiventer*" (p. 100). They leave the question unresolved, and assign no specific name to their specimens (although they appear conspecific with *N. masalai*). As well as cranial and dental features, our specimens share with *N. cephalotes* a sexual dimorphism of color (males grayer, females paler brown) (Andersen 1912). Smith and Hood (1981) originally listed *N. cephalotes* from New Ireland, and we assume that they later ascribed these to *N. vizcaccia* (a similar form), or to the new species, *N. masalai*, because neither they (1983) nor Flannery and White (1991) later list *N. cephalotes* from New Ireland. Specimens in the USNM assigned to *N. albiventer* from the Moluccas are much smaller, with distinctive, anterior-posteriorly slender canines that bear a large posterior cusp.

[2]Taxonomy follows Griffiths et al. 1991. One specimen (USNM580037) seems typical of *M. nigrescens* papuanus in size and all cranial and external features. Other specimens (USNM580034, 580068, LHE1047, LHE1050?) are a different color, larger than typical for *M. nigrescens*, have some cranial differences, but have the penis, tragus, and some cranial features of *Mosia*. They are virtually identical to an *M. "nigrescens"* from St. Matthias Island (USNM 277113). Koopman (1979) noted that *Mosia nigrescens solomonis* from New Britain was larger and distinct from *M. n. papuana* from New Guinea. We tentatively assign our larger *Mosia* to *M. n. solomonis*, and the smaller to *M. n. papuana,* and because they are sympatric, they each deserve species status.